Biological Life Support Systems

Edited by

Mark Nelson
Space Biospheres Ventures
Oracle, Arizona

Gerald Soffen
Associate Director of Earth Studies Directorate
NASA Goddard Spaceflight Center

Proceedings of the Workshop on Biological Life Support Technologies: Commercial Opportunities sponsored by the National Aeronautics and Space Administration Office of Commercial Programs (Technology Utilization), Washington D.C., hosted by Space Biospheres Ventures, and held at the Biosphere 2 Project, Oracle, Arizona, 1989

Space Biospheres Ventures

SYNERGETIC PRESS

SP

Published by Synergetic Press, Inc.
    Post Office Box 689
    Oracle, Arizona 85623

These proceedings were originally published by (and reprinted with permission of) the National Aeronautics and Space Administration, Office of Management, Scientific and Technical Information Division, as a NASA Conference Publication, under the title *Workshop on Biological Life Support Technologies: Commercial Opportunities,* Mark Nelson and Gerald Soffen, editors.

Production and Design Editor: Kathleen A. Dyhr

Cover design: Kathleen A. Dyhr

Copy Editor: Scott McMullen

Front cover photograph: Space Biospheres Ventures Biosphere 2 Test Module, Oracle, Arizona.

Front cover photograph is copyright © 1988 by Roger Ressmeyer—Starlight, and originally appeared in *National Geographic* magazine.

ISBN 0 907791 22 0

Printed in the United States of America by Arizona Lithographers, Tucson

Contents

Foreword

A workshop on Biological Life Support Technologies: Commercial Opportunities — sponsored by the NASA Office of Commercial Programs and hosted by Space Biospheres Ventures — was held at the Biosphere 2 project site near Tucson, Arizona from October 30 to November 1, 1989.

The meeting drew together researchers from some of the most innovative projects of NASA Life Sciences and private industry in environmental and bioregenerative systems. The role of biological life support technologies was evaluated in the context of the global environmental challenge on Earth and the Space Exploration Initiative, with its goals of permanent space station, lunar base and Mars exploration.

Background on Biological Life Support Systems Research

Research on biologically-based systems of life support dates back to the 1950s spurred by the advent of high altitude flight and the development of submarines where crew members spent long periods of time in tightly sealed environments, separated from the normal life support mechanisms provided by the biosphere of the Earth. The beginning of space flights greatly accelerated this research, with programs supported by both the NASA and Soviet space agencies.

The driving force behind the search for biological — also called "bioregenerative" — life support systems derives from the implications of a simple calculation. Approximately 0.6 kg food, 0.9 kg oxygen, 1.8 kg of drinking water, 2.3 kg of sanitary water and 16 kg of domestic water for a total of some 22 kg per day, or some 45-50 pounds are required to provide life support for each person for every day in an artificial life support system. Thus, in the course of a year, the average person consumes three times his body weight in food, four times his weight in oxygen, and eight times his weight in drinking water. Over the course of a lifetime, these materials would amount to over one thousand times an adult's weight.

Life support systems for astronauts have been accomplished to date by almost purely physico-chemical means and mainly by supply from Earth. When provisions of food, water and breathable atmosphere are consumed, astronauts must return to Earth — or be resupplied, as in the Soviet *Mir* Space Station, by expensive materials relaunched from the bottom of Earth's gravity well. The role of a sustainable biologically recycling system will be to radically change our ability to sustain human life in space on a permanent and evolving basis.

The early search for developing biological life support concentrated on aquatic tanks for growing highly productive algae for regenerating air and water. Despite intense efforts, it proved impossible to make the algae palatable to humans except in quantities of 25-50 grams per day. It also proved difficult to couple the algae reactors to higher plants in integrated life support systems because byproducts of the algae are injurious to plants.

Nevertheless, research has continued, now focussing on systems based on higher plants for food production. The Soviet program included the Bios-3 facility at the Institute of Biophysics, Krasnoyarsk, Siberia where crews of two to three people were supported for up to six months in a biologically-based system where nearly all of the water, 95 percent of the air and about half of their food was produced/regenerated within the facility. In 1978, NASA initiated its Controlled Environmental Life Support Systems (CELSS) program to develop capability in biological life support and in 1984, Space Biospheres Ventures started its program of bioregenerative life systems research and development.

Applications and Commercial Opportunities

Biological life support technologies have a spectrum of terrestrial applications in addition to their role in space exploration and habitation. Some of the problems that they address have strong counterparts in problems of environmental cleanup and recycling. For example, the problem of clean air regeneration in spacecraft cabins or sealed space outposts is analogous to the so-called "sick building syndrome" often associated with relatively tightly sealed energy-efficient buildings. A static atmosphere — without a means of removing trace gases from outgassing of materials and from people themselves — eventually leads to accumulation of trace gases and possible health problems. The challenge to recycle human waste and regenerate domestic/hygiene water in space habitats is similar to the problem of developing recycling systems in the global arena to prevent the degradation of waters caused by urban sewage disposal.

The goal of developing bioregenerative life support systems which completely sustain humans in a clean and healthy atmosphere — and which do not produce pollution as a byproduct — has parallels with humanity's need to create technologies that will permit development without eroding the habitability or life support capacity of our planetary home.

The commercial opportunities and historic importance for such spinoffs are enormous as we begin to cleanup past pollution and develop non-polluting types of technology. In addition, development of completely materially closed life systems opens a wealth of scientific opportunities. Elucidation of the biotic cycling exhibited in such systems can lead to fundamental insight into how such processes operate in natural ecosystems. Small, materially-closed systems — such as those being developed by the NASA CELSS program, and biospheric and Test Module-type laboratories by Space Biospheres Ventures — have short cycle times for nutrients and gases and allow very intensive monitoring, thus facilitating detailed examination of mechanisms of interest to scientists studying Earth's geosphere and biosphere. We have begun to study the Earth as a total system from space, and will increasingly do so in the coming decades under the various "Mission to Planet Earth" programs of space agencies, and the International Geosphere/Biosphere Program of ICSU.

Review of Workshop

It was in this context of expanding possibilities and necessities that the workshop was convened.

Dr. Thomas Paine, former Administrator of NASA and Chairman of the National Commission on Space, opened the workshop with a tour of space frontiers which beckon within our extended environment, the Solar System. Dr. Paine underlined the role and necessity for bioregenerative and biospheric systems to make our exploration and habitation permanent.

John Allen, Director of Research and Development at Space Biospheres Ventures (SBV), presented a historical overview of the Biosphere 2 project, underscoring the potential of closed biospheric systems to advance understanding and stewardship of the global biosphere on Earth. A Biosphere 2 project site tour provided background and introduction for closed ecological systems presentations by SBV. Abigail Alling, SBV Director of Marine Ecological Systems, reviewed the Biosphere 2 Test Module manned and unmanned closed ecological system experiment series, as well as engineering innovations used to produce such a tightly sealed structure. She outlined Test Module-tested subsystems and technologies which have for the first time provided complete bioregenerative life support. Carl Hodges, Director, and Dr. Robert Frye of the Environmental Research Laboratory, detailed soil microbial air purification technology utilized in intensive food production systems for Biosphere 2, noting potential spin-offs for sustainable high yield and non-polluting agricultural systems and for solving air quality problems. Dr. Roy Walford, Professor of Pathology at UCLA Medical School and SBV Chief of Medical Operations, outlined the health program for Biosphere 2, major potential concerns and monitoring procedures, including use of micro-systems for diagnostic tests and biomarker studies for long term evaluation of human health in closed ecological systems.

Lee Tilton introduced the environmental engineering and biological life support research of NASA Stennis Space Center, where he is Director of Science. Dr. Bill Wolverton, now of Wolverton Environmental Consulting and recently retired from NASA Stennis, discussed the natural ability of plants and their microbial symbionts to solve waste recycling and pollution problems of soil, air and water, emphasizing the enormous economic and environmental potential for such biological systems. Microbiologist Anne Johnson presented the BioHome — a practical application of Stennis Space Center Environmental Laboratory research — integrating biological waste, water and air recycling systems into a private home.

Dr. Mel Averner, Manager of the NASA Controlled Environmental Life Support Systems (CELSS) and Biospherics Programs, introduced the NASA CELSS program, its major drivers and current requirement to be able to evolve from a "hybrid system", interfacing with partly physico-chemical life support technologies. Dr. Bill Knott, Manager of the Life Sciences Support Facility at NASA Kennedy Space Center, reviewed the most advanced CELSS endeavor to date, the Breadboard Project —a Project Mercury pressure chamber reconfigured as a biomass production unit with a closed and recycling air and water supply. Dr. Knott outlined its control and monitoring systems, the dynamics of environmental parameters and its record of production with recent cropping trials. Dr. David Bubenheim, Research Scientist at the NASA Ames Research Center, discussed CELSS work at Ames to improve system efficiency, and the Plant Growth Research Chamber as a prototype for bioregenerative systems flight hardware.

Dr. Gerald Soffen, Associate Director for Program Planning and Chief Scientist of Earth Observation System (EOS), NASA Goddard Spaceflight Center, took us on a Mission to Planet Earth. He reviewed the urgent necessity for understanding the Earth as a system because of human impact on the global environment. He reviewed the EOS program as an example of international cooperative ventures where space platforms and terrestrial studies can cooperate to revolutionize our knowledge of the biosphere.

Dr. Joseph P. Allen, President of Space Industries Inc. and former NASA Space Shuttle astronaut, contributed a first hand report of space life support to date — from the shuttle cabin to free-falling in the life cocoon of a space suit. He voiced support for commercial initiatives as well as new and less cumbersome management approaches to enable both NASA and the private sector to better bring space opportunities and visions to fruition.

Dr. Wendell Mendell, Chief Lunar Scientist, NASA Johnson Space Center, reviewed the emergence of a practical strategy for the evolutionary expansion of humankind into the Solar System. He pointed out that a chief limiting factor now in such plans, which are now official U.S. and NASA policy, is our understanding of life sciences in space and especially bioregenerative life support technologies. He presented a new trade-off study on the payback period for bioregenerative life support and the historic opportunities now available since the policy decision to establish a permanent lunar outpost as a step toward Mars exploration and bases.

Mark Nelson,
Gerald Soffen

Acknowledgements

Grateful acknowledgement is due to Ray Gilbert of the NASA Office of Commercial Programs (Technology Utilization) and Dr. John Cleland of the Research Triangle Institute for initiating and helping ensure the success of the workshop; and to Dr. Mel Averner who organized and ably chaired the NASA Life Sciences presentations.

Biospheres and Solar System Exploration

Thomas O. Paine, Ph.D.
Administrator of NASA, 1968-70
Chairman, National Commission on Space, 1984-86

I believe in the research program initiated here at Space Biospheres Ventures. Humanity is a flourishing species because of our drive to explore and our technological ingenuity. Twenty thousand years ago our ancestors initiated the agricultural revolution with technologies that altered our relationship to nature. Herdsmen and cultivators can't revert to hunter-gatherers, nor can we abandon our half-completed industrial revolution, although we must better manage the environmental impact.

We meet here at a time of historic decision with modern nations at a crossroads, reconsidering the choice between developing technology for mutually-assured destruction, or for expanding life beyond Earth's biosphere. The news in late 1989 is encouraging. The superpowers appear to be turning onto the road of life, but — human nature being what it is — the issue can never be finally resolved. The next great evolutionary challenge to our species is to open the Inner Solar System to human settlement. Learning to "live off the land" on resource-rich Mars will double the territory available for life, and encourage our descendants in another century to settle countless planets circling distant suns.

In this historical context, I see Biosphere 2 as a shining beacon pointing the way to an expanding future for humanity. Closed ecology systems can free us from Malthusian limitations by making the Solar System our extended home. For the first time in the history of evolution, the human intellect can extend life beyond Earth's biosphere, following the lead of species that left the oceanic biosphere to inhabit dry land billions of years ago. In the 21st Century, a network of bases throughout the Inner Solar System, interconnected by space transportation and communication infrastructure, can sustain vigorous high-tech civilizations evolving on three worlds. The space settlement implication of Biosphere 2 is thus my theme for tonight.

IMPLICATIONS OF BIOSPHERE TECHNOLOGY

As you know, our Earth is one of nine known planets circling the Sun, which is one of about a trillion stars in our Milky Way Galaxy, which is one of about a trillion observable galaxies (which will probably grow to ten trillion galaxies when the space telescope goes into operation next year). So we have explored only eight of the universe's trillion trillion terrae incognitae (Figure 1). We can't snap a photo of our own Galaxy, but we can photograph the nearby Andromeda galaxy, which closely resembles our Milky Way. Our Sun is a star out near the galactic rim; it is from this perspective that we observe the heavens.

As far as science can tell, the only life in the entire cosmos is that riding through space on our precious blue planet, and the only intellect in all creation studying the universe is the human brain. With a trillion trillion possibilities, it's hard to believe that we're alone, but to date we have turned up no scientific evidence for the existence of life beyond Earth. So we are "E.T." — it's up to us to expand intelligent life to the stars.

THE FIRST STEP: OUR SOLAR SYSTEM

Our energy-giving Sun is circled by the four earth-like planets: Mercury, Venus, Earth, and Mars. Beyond Mars lies the asteroid belt, where more than 3000 small planetesimals have been discovered (and more than ten times that number are believed to exist). Beyond these are the four gas giants of the outer solar system: Jupiter, Saturn, Uranus, and Neptune, then the outermost planet, Pluto, and finally the great Oort cloud of comets extending for billions of miles. Occasionally one of these icy bodies is perturbed and swings through the Inner Solar System, boiling off a vaporous tail which the solar wind deflects across the night sky. Cometary impacts may have distributed water and organic chemicals throughout the Solar System from the enormous quantities stored in the Oort cloud. Let's briefly review the exploration status and prospects of each world in our Solar System.

The Sun

As Copernicus and Galileo showed, our Sun is the central star whose thermonuclear cycle provides the life-giving energy that drives Earth's biosphere. Surprisingly, we still don't fully understand the nuclear fusion cycles involved; the Sun's neutrino flux doesn't quite fit our physics equations. Since the Sun fuels all life, space-based observatories and underground neutrino detectors are being improved to clear up the mystery of solar physics.

Figure 1. A portrait of seven of the planets in our Solar System studied by spacecraft. The one in the middle, Earth, has a unique life support system called a biosphere. (Photo: NASA.)

Mercury

The planet nearest the sun is slightly larger than our Moon. Mercury's surface resembles the Moon's because both sustained intense meteorite bombardment in the early history of the solar system, and an absence of water erosion preserved their cratered terrains. The prospects of astronauts exploring Mercury soon are remote. Because it is so close to the sun, elaborate thermal protection would be required on the illuminated side. On the other hand, we and the Soviets are discussing automated Mercury probes early in the next century, and it is certainly possible that humans might explore the planet later if sturdy robots find interesting resources and research opportunities.

Venus

Next comes Earth's twin planet, Venus, with its dense atmosphere of carbon dioxide, sulfuric acid, and other gasses. Atmospheric scientists have a fine laboratory here in which to study a run-away greenhouse effect. The pressure at the surface of Venus equals that two thousand feet beneath the ocean, with a temperature high enough to sustain puddles of molten lead. U.S. and Soviet spacecraft have shattered science fiction dreams of humid jungles teeming with seductive Amazons. Cloud-shrouded Venus has been mapped by orbiting side-looking radars, and several Soviet landers have parachuted to the hostile surface to transmit brief observations of basaltic rocks before being

Figure 2. Mars settlement in the 21st century. In the distance, a spacecraft departs the Martian base. (Artist: Robert McCall. Copyright 1986 by Bantam Books, Inc.)

incinerated. Powerful radar signals bounced off Venus from our giant Arecibo radio telescope in Puerto Rico show shiny areas at the base of conical peaks, suggesting major flows of volcanic lava. NASA's Magellan probe is now en route to Venus to obtain a high precision map of the Venusian mountains and plains.

Earth

Next outward from the Sun is our own beautiful blue planet, 75% covered by oceans. Distant photographs by Apollo astronauts of Earth's unique biosphere floating in space provided great impetus to the environmental movement. Space observations allow us to scan continuously the entire surface of Earth, monitoring ozone, agriculture, glaciers, tectonic plates, polar icecaps, vulcanism, the interaction of ice and water with the atmosphere and land, and many other critical processes. From orbit we can study pollution and urbanization, the destruction of great rain forests, desertification, erosion, and resulting changes in the Earth's climate.

In 1992 a major Mission to Planet Earth will celebrate the 500th anniversary of Columbus' discovery of a new world. Many nations will join an intensive Earth monitoring program combining space and surface systems. Photographs from space will record the temperature of the entire globe each day of the year, while other satellites scan auroral zones. When I visit my Alaskan daughter and watch the beautiful northern lights, I can't see that the flickering sheets of solar ions extend all the way around the magnetic pole, but satellites can. The pioneering flight of the Wright brothers reminds us that the most interesting phenomenon on our planet is the human intellect. Sixty five years after the first airplane flight in 1903, Apollo astronauts flew 240,000 miles from Earth to explore the Moon.

The Moon

Although the barren lunar surface provides a great contrast to Earth's teeming life, we've operated six research stations there, and a dozen astronauts have traversed the cratered terrain. The Moon is a geologist's paradise of ancient rock formations.

We've learned a great deal about lunar resources from the Apollo expeditions. The rocks are about 40% oxygen, which can be extracted for life support and spacecraft propellants. Terrestrial plants thrive in lunar soils, which contain finely powdered glasses, metallic particles and minerals. Indigenous resources will be valuable for future lunar operations, including a rich inventory of heavy elements, but the Moon lacks water. Hydrogen, carbon, and other essential light elements are scarce on the Moon, but abundant on Mars.

Mars

Humanity's next destination in space is resource-rich Mars and its moons (Figures 2 and 3). Voyaging hundreds of times the lunar distance from Earth will become routine in the first quarter of the 21st Century. Robotic spacecraft orbiting Mars have transmitted detailed photos, including spectacular features like Mons Olympus, the greatest volcano in the solar system. This giant cone spreads 420 miles across the plain and soars 15 miles to a lofty caldera. The enormous bulk reflects the lack of tectonic plate movement on Mars. We believe that the Hawaiian Islands were formed as a tectonic plate drifted above a subterranean magma source, throwing up a long chain of volcanic islands. On Mars, however, the plates appear to be fixed, so volcanoes grew larger and larger. This is just one of many terrestrial insights scientists are gaining from comparative planetology.

The most surprising discoveries from Viking spacecraft orbiting Mars were pictures showing evidence that at one time liquids flowed across the Martian surface. No rivers can exist today because the pressure of the thin carbon dioxide atmosphere is below the critical point of water; Martian ice therefore sublimes directly into vapor. Yet water eroded the surface for some time after the Martian impact craters were formed, and underground permafrost may still exist. Further evidence is provided by impact craters that show a muddy-looking fringe, as though the heat of collision produced a mushy outward wash. Looking down from orbit in the early morning we saw water fog forming in some valley areas, so substantial water resources exist in the atmosphere. Martian water frozen in polar icecaps, possibly underground, and in the atmosphere, will provide future pioneers with a resource essential for life.

Two robotic Viking explorers landed on Mars in 1976 carrying TV cameras, weather stations, and life-detection experiments. Their transmitted data followed the seasons throughout the Martian year (669 24-hour, 40-minute days), including great planet-wide dust storms. Pictures they took of a frosty morning on Mars shows the abundance of extractable water in the atmosphere. The soils sampled revealed no organic materials or evidence of life. Although these results were negative, life may exist elsewhere on Mars. The era of liquid water on Mars lasted longer than the time required for the first terrestrial life to appear in Earth's oceans, so fossils may record earlier life. We have much yet to learn about the possibility of life beyond Earth, and Mars is a superb laboratory.

Asteroid Belt

Beyond Mars lies a swarm of small asteroids that never aggregated to form a planet, but remain as tens of thousands of planetesimals. The Martian moons, Phobos and Deimos, are believed to be captured asteroids. As NASA's Galileo spacecraft flies through the asteroid belt on its six-year journey to Jupiter, it will observe asteroids Gaspra in Octo-

Figure 3. Mining propellant on Phobos, a moon of Mars. (Artist: Robert McCall. Copyright 1986 by Bantam Books, Inc.)

Biological Life Support Systems

ber, 1991, and Ida in August, 1993. All future planetary missions beyond Mars will be targeted to fly by asteroids. In the 21st Century, six-month piloted missions to nearby asteroids should follow the initial human exploration of Mars.

Jupiter

Beyond the asteroid belt is giant Jupiter, which contains most of the mass of the solar system outside the Sun. One of its remarkable moons is Io, with active volcanoes that spout sulfur high into the sky. These volcanoes were actually discovered by a computer and an alert technician, Linda Morabito, of NASA's Jet Propulsion Laboratory. She fed incoming photos into an automated navigation program that pinpointed spacecraft position by scanning the limb of the moon in relation to nearby stars. When the computer kept rejecting the pictures of Io's limb, she checked and noticed a mushroom cloud where no cloud should be. Additional pictures showed soaring volcanic plumes distorting the smooth arc of Io's horizon; thus, to everyone's amazement, vulcanism was discovered in the Outer Solar System.

Other Jovian moons show intriguing features, too, including Ganymede, Callisto, and Europa, with ice-crusted oceans. The Galileo spacecraft will study them all after it deploys a European Space Agency probe into Jupiter's atmosphere. The isotopic compositions of Jupiter's gasses is of great interest to planetologists and astrophysicists, since they preserve the primitive material from which the Solar System was born. Galileo will

Figure 4. In the foreground is an aerospace plane and the Earth Spaceport. The spaceport is receiving cargo from a cargo transport vehicle (lower left-hand corner). In the background, a two stage transfer vehicle is returning to the Earth Spaceport from the Moon. (Artist: Robert McCall. Copyright 1986 by Bantam Books, Inc.)

journey throughout the Jovian moon system for several years, transmitting back to Earth pictures with a thousand times the resolution of previous images.

Saturn

Next after Jupiter is spectacular Saturn, with its magnificent ring structure. In April, 1996, NASA plans to launch the Cassini Mission with a European lander targeted for Titan, Saturn's largest moon. Titan's cloudy atmosphere is rich in organic compounds, which react under solar and cosmic ray irradiation to form Los Angeles-like smogs. From Titan's clouds methane may snow onto oceans and glaciers of organic compounds and continents of ice. The European Space Agency's probe should give us an exciting view of Titan's surface, perhaps shedding light on the conditions on ancient Earth when organic molecules first combined to form living systems. After deploying the ESA probe at Titan, the Cassini spacecraft will carry out an ambitious observation program of Saturnian moons and rings.

Uranus and Neptune

Beyond Saturn is Uranus, with equally fascinating moons. Miranda, for example, appears to have suffered an enormous impact that fractured it into a number of fragments, which then reconstituted themselves under low gravity into an incredibly jumbled topography. The geology of the Uranian moons, and of the more distant Neptunian moons, exhibit a fascinating diversity. Voyager's final detailed photographs of the large Neptunian moon, Triton, show geysers spouting liquid nitrogen five miles into the atmosphere, with debris falling onto continents of ice along a line 60 miles downwind. Triton is indeed a fascinating world.

Pluto

The one planet NASA's far-ranging reconnaissance robots have yet to visit is remote Pluto and it's large moon, Charon. This distant duo also promises to exhibit the diversity we've come to expect in the outer Solar System. We need to understand the energetics of these worlds far from the Sun, which appear to emit more energy than they receive. Missions to Pluto/Charon involving flight times up to forty years are under study. NASA's reconnaissance of our Sun's planets and moons is teaching us much that is applicable to potentially habitable worlds circling other stars.

Comets

Comets bring primitive material from the fringes of the Solar System into range of our spacecraft. Early in the next century NASA is planning to land a probe on a comet as it passes the orbit of Jupiter on its inward journey past the Sun. The goal is to monitor the comet through its closest approach to the Sun, studying the emissions from its outgassing surface as they stream out to form the tail. A refrigerated sample of its icy core may be brought back to Earth for study by a parallel probe that the Japanese Space Agency and NASA are discussing. Such a sample would represent invaluable material from interstellar space.

THE COMING EXTRATERRESTRIAL CENTURY

The 21st century will usher in a new Age of Discovery based upon reliable, low cost travel throughout the Inner Solar System. President Bush has directed his National Space Council and NASA to prepare plans for an evolutionary space station in Earth orbit in the next decade, a return to the Moon to establish permanent bases about 2004, and the manned exploration of Mars starting about 2015. This follows the recommendations of the National Commission on Space Report, Pioneering the Space Frontier[1], which listed five program elements as particularly critical for future interplanetary operations:

1. A Highway to Space to provide reliable, low-cost access to Earth orbit for passengers and cargo;

2. Orbital Spaceports circling the Earth, the Moon and Mars, to support spacecraft assembly, storage, repair, maintenance, refueling,

check-out, launch and recovery of robotic and piloted spacecraft:

3. A Bridge between Worlds to transport cargo and crews to the Moon, and to extend human spaceflight hundreds of times the lunar distance to Mars, with cycling spaceships in permanent orbits between Earth and Mars;

4. Prospecting & Resource Utilization Systems to map and characterize the resources of planets, moons and asteroids, and learn how to "live off the land" using indigenous materials on other worlds; and

5. Closed-Ecology Biospheres, like Biosphere 2, that can provide food and recycled air and water within secure habitats remote from Earth.

Each of these five elements is challenging, and each requires technological advances across a broad front. Yet we know much more today about establishing a network of evolutionary outposts and bases around the Inner Solar System than we knew about lunar landing when President Kennedy initiated the Apollo Program in 1961. We also have a broader base of international cooperation, a larger gross world product, and far greater astronautical experience. Let's review progress in each of these five fields.

The Highway to Space

Our most urgent need is a significant reduction in the cost of transporting cargo and crews between Earth and Low Earth Orbit. The U.S. Space Shuttle pioneered high-pressure hydrogen/oxygen engines, recoverable solid boosters, lightweight structures, high temperature re-entry tiles, automated landing from orbit, winged flight through the range of Mach numbers from zero to twenty five, vehicle reusability, payload return to Earth, and many other significant innovations. It is a superb craft for carrying 2 to 8 astronauts and substantial payloads between Earth and orbit in infrequent missions lasting several weeks. But the objective of routine low-cost transport cannot be achieved by this piloted vehicle, and shuttle operations are too expensive to continue indefinitely. Candidate new piloted systems include the Advanced Launch Sys-

tem (ALS), a Personnel Launch System (PLS), and the X-30 National Aero-Space Plane (NASP). The shuttle has taught us much about the system requirements for routine access to orbit, but a major reduction is needed in the cost of transporting large tonnages of cargo into orbit for 21st Century operations on the Moon and Mars.

Commercial cargo launch services are now available from many nations, but most employ labor-intensive, one-shot, missile technologies from the 1960s, with inherent high cost and single-point failure modes. New Ariane, ALS, and other launch vehicles are in prospect, but launch technology is about where aircraft design was in the 1920s, when barnstorming pilots flew with canvas and piano wire. But we can envision a future space transport equivalent of the economical Douglas DC-3, and the required technology base is under development in NASA's "Civil Space Technology Initiative" and "Pathfinder Program" (R&D in support of Solar System exploration).

For cargo transport, NASA is studying an unmanned Shuttle C, and a joint NASA-Air Force Advanced Launch System. Similar programs are under study by other countries. Now that President Bush has set the long-range U.S. goal of exploring Mars via the Moon, NASA can specify the characteristics of future payloads and launch systems. Serial production of fully-automated launch vehicles will significantly reduce the cost and hazards of spaceflight.

Orbital Spaceports

The U.S. *Skylab* and U.S.S.R. *Salyut* and *Mir* space stations have demonstrated the feasibility and utility of manned orbital laboratories. Astronauts and cosmonauts have carried out Earth observations, zero-gravity processing, ultraviolet and X-ray astrophysics, studies of the physiological effects of months of prolonged weightlessness, and many other experiments. Cosmonauts aboard the space station Mir have conducted medical and biological experiments demonstrating the possibility of a one-year, zero-g flight to Mars. Modules for the international Space Station *Freedom* are being designed by NASA and the European Space

Agency (ESA) in collaboration with Japanese and Canadian teams.

The new challenge is to design Space Station *Freedom* for the mid-90s with the flexibility to evolve into an international Spaceport by the turn of the century. Spaceport Earth must also provide the prototype for Spaceport Moon by 2001, and Spaceport Mars a decade later. This will establish an international network of orbital bases around the Inner Solar System combining the functions of space transportation nodes, communication centers, space laboratories, habitats, medical outposts, general purpose workshops, spacecraft assembly and checkout facilities, supply depots, maintenance bases, and fuel farms. Just as seaports assemble and service ships, orbital spaceports will assemble and service spacecraft. They will support a diverse fleet of satellite platforms circling three worlds, and dispatch and recover spacecraft for interplanetary cargo and passenger transport (Figure 4).

The Bridge Between Worlds

Modular space transfer vehicles with hydrogen-oxygen engines and aerobraking shields are being developed for Earth-Moon and Earth-Mars cargo and passenger flights. Lower-cost cargo transport is in prospect using low-thrust, high-specific-impulse solar or nuclear electric propulsion systems, with the propulsion electric generators adding to the useful delivered payloads.

Large cycling spaceships swinging permanently between the orbits of Earth and Mars appear promising for interplanetary passenger transport in the 21st Century. Aerobraking transfer vehicles can ferry passengers between the cyclers and Spaceport Earth at one end of the voyage, and Spaceport Mars at the other, eliminating the need to accelerate and decelerate the massive transports. Cycling spaceships on the Mars run will be more like the *Queen Elizabeth II* than a Boeing 747. Their large mass and volume will permit redundant power and life-support systems, well-equipped laboratories, comfortable living quarters, and closed-ecology biospheres (future generations of Biosphere 2). Safety features will include heavy shielding to pro-

tect crews from cosmic rays and solar flares, medical clinics, artificial gravity chambers, exercise gyms, and other health maintenance facilities.

Although Apollo demonstrated the feasibility of expendable spacecraft for flights to the Moon, NASA's new Martian goal suggests using prototype cycling spaceships on the Earth-Moon run to gain operational experience. Well-equipped lunar cyclers would also give scientists valuable research platforms for interferometry and other deep space experiments, and allow engineers to check out robotic operation, artificial gravity chambers, and closed-ecology biospheres with 24 hour daily illumination. During solar flares and passages through the Van Allen Radiation Belts, lunar travelers would be protected by the massive shielding that will be required aboard spaceships on the Mars run.

Prospecting and Resource Utilization Systems

Automated and piloted lunar orbiters, landers and rovers have taught us much about the Moon's resources, but we've literally only scratched the surface. The scarcity of hydrogen and other light elements on the Moon may make it less promising than Mars for self-sufficient settlements in the long run, since water may have to be imported. Sunless craters at the lunar poles might contain trapped volatiles like ice, however, so polar prospecting is planned in the next few years, starting with a Japanese lunar probe. The Moon's proximity to Earth permits teleoperated systems, which are difficult on Mars due to communication time delays across tens of million of miles. Robotic mapping, prospecting, and sample-return rover missions in the next decade will provide the engineering data needed to design Lunar and Martian bases.

Over the next 40 years, we must develop the broad technology base, transportation infrastructure, and network of self- sustaining bases beyond Earth that will permit men and women to "live off the land" on the space frontier. In addition to habitats and laboratories, Lunar and Martian bases will require solar or nuclear electric generators in the 1-10 megawatt range, automated plants to process indigenous materials, construction machinery, general purpose robotic fabrication plants (with software links to twin factories on Earth), maintenance shops, and transportation support facilities. Innovative architecture should take advantage of the Martian environment; for example, on-site materials with an ice binder can substitute for concrete on sub-freezing Mars. With NASA's sights set ultimately on Mars, Lunar base prototype systems should be specifically designed for adaptibility to Martian conditions.

Closed-Ecology Biospheres

To support people living in bases remote from Earth, air and water must be recycled, and nourishing food produced within automated, closed-cycle life-support systems like Biosphere 2. Air and water have been successfully regenerated in prototype systems, and the problems are reasonably well understood, but little is known about constructing reliable biospheres that can be depended upon to supply food and fiber. Closed-ecology experiments include the Soviet Bios-3 project and NASA's Closed Ecology Life Support Systems (CELSS) projects. Test subjects have spent more than six months sealed within Bios-3, although some food was imported. Less ambitious, but more compact, closed-ecology systems are being studied at NASA's Kennedy and Johnson Space Centers. Of all the critical elements, the Space Biospheres Ventures' goal of a closed-ecology biosphere remains the least understood and the most challenging, so you can understand why I'm enthusiastic about Biosphere 2.

CONCLUSIONS

Scientific progress interacting with the vastness of the Space Frontier can eliminate Malthusian limits to human aspirations. Our advancing technology base is ushering in an age of space exploration that has already brought great rewards to Earth, and in the 21st Century will expand life from its earthly cradle to the Moon and Mars. Developing the limitless space frontier will contribute to science, technology, productivity, economic growth, education,

medicine, agriculture, international cooperation — indeed, to every feature of terrestrial life. Beating our terrestrial swords into extraterrestrial plowshares can convert yesterday's arms race into tomorrow's international space settlement.

Establishing a base on Mars and supporting it will be well within our capabilities by 2015. As I've stressed: the critical problem is learning to "live off the land" on Mars. Since we can't carry frozen dinners from Earth across millions of miles to Mars, Biosphere 2 is essential to make Mars self-supporting.

What about the distant future? Let me close by considering the Drake Equation, which starts with the trillion suns in the galaxy and the trillion galaxies in the universe, and estimates the probability of life beyond Earth. For a first approximation, multiply the number of stars formed each year, times the fraction of the stars that have planets, times the fraction of planets where water is liquid, times the fraction where life develops, times the fraction with evolutionary species, times the fraction with intelligent beings, times the fraction that develops technology, times the fraction that wishes to communicate across the cosmos before they wipe themselves out or lose interest. Despite all these fractions, you begin with such large numbers that it appears life must exist elsewhere. So the search has started; the Planetary Society, NASA, Soviet observatories, and others are operating banks of computers linked to large antennas that scan the sky for an E.T. "I Love Lucy" broadcast.

We do have initial evidence for the existence of planets in other solar systems. Recent observations of Beta Pictorus by an infrared satellite show material in orbit around the star. This evidence, and the history of our own Solar System, suggests that planets may be the usual result of star formation. We still don't know whether life normally appears and evolves on temperate aqueous planets. As M.I.T.'s Philip Morrison points out, however: either there is life elsewhere in the universe, or there is not — and in either case it boggles the mind!

If we can detect planets circling a nearby star, using observatories in Earth orbit or large infrared telescopes on the back of the Moon, and if one of them exhibits an atmospheric spectrum showing the presence of water vapor and plant-generated oxygen, I'm sure that our grandchildren or great grandchildren will organize a new megaproject to dispatch a starship across light-years of interstellar space. We may not live to see that, but we saw Apollo astronauts launch the exploration of other worlds (Figure 5). The Biosphere 2 Project is contributing to the critical next step: closed ecology systems that will expand terrestrial life throughout the Solar System.

REFERENCES

1. *Pioneering the Space Frontier: Final Report of the National Commission on Space,* Bantam Books, New York 1986.

Figure 5. Apollo 16 astronaut Charles M. Duke Jr. collects lunar rock samples, April, 1972. On the lunar surface, Duke and John Young collected over 200 pounds of rock samples including one determined to be 4.25 billion years old, thought to be part of the Moon's original crust. (Photo: NASA.)

Historical Overview of the Biosphere 2 Project

John P. Allen
Director of Research and Development
Space Biospheres Ventures

In late 1969, as the moon landing commenced, the Institute of Ecotechnics also started, at first on a very small scale, to work on ecological projects which laid the conceptual foundation for the current Biosphere 2 project. These projects were designed to bring together ecological and scientific knowledge with appropriate technics to design economically viable and ecologically-upgraded total systems in a spectrum of challenging biomes around the world. The Institute of Ecotechnics was motivated to begin this line of research and development because, as Tom Paine noted, we could see that biospherics is one of the key scientific fields we have to master for life to succeed both on and off the planet.

Of course, the space program was an important ingredient in giving a new impetus to biospherics. The Russian scientific tradition is quite interesting in the equal emphasis it gives to Konstantin Tsiolkovsky, a founder of astronautics, and Vladimir Vernadsky, who laid the scientific basis for understanding the biosphere. Tsiolkovsky developed, along with Goddard in our country, the practical foundations of the idea of rocketing into space. Vernadsky pointed out that life itself is a tremendously powerful geological force, far more than the common perception of it as a thin shell surrounding a small planet. He saw life and the biosphere as a cosmic phenomenon, both because it fundamentally depended on cosmic energy coming in — solar radiation — and because it was an immensely powerful force that could transform the surfaces of planets. Vernadsky came to the same conclusion as Tsiolkovsky, namely that biospheres

were destined to go into space, outgrowing their planetary cradle here on Earth. By 1969 the famous photographs of the blue planet seen from space had begun to change the way all of us thought and felt about the Earth, leading to a flowering of studies of planetary ecology. G.E. Hutchinson of Yale, who was a great American student of Vernadsky, edited the influential Scientific American volume The Biosphere published in 1970. But still many questions remained. How, actually, could you put a conceptual model of Earth's biosphere together, containing and regulating as it does such vast, marvelous and evolving complex systems?

About that same time, in 1968, Clair Folsome, who had consulted to NASA on the origins of life and was Director of the Exobiology Laboratory, University of Hawaii at Manoa, took a complete functional suite of microbes together with their associated aquatic element and an air volume and put them inside a closed laboratory flask in which he could measure the oxygen and carbon dioxide levels, study energy flows and visually observe changes. For the first time there was a closed ecological system object for scientific study. These closed laboratory ecospheres prove to be indefinitely viable and regenerating given an energy input as long as a sufficiently diverse functional complement of microbes is enclosed. The 1968 flask with its living ecosphere is among the collection of Clair's laboratory systems maintained for their historical and continuing research interest in the Space Biospheres Ventures Analytical Laboratory building. Clair, who had served on the Biosphere 2 Project Review Committee since its beginning, died unexpectedly last year. His work and vision continues at our Biospheric Research and Development Center and at many laboratories which continue working on the dynamics of closed ecological systems. Clair's research showed that each of these "worlds" establishes its own gas/water balance and metabolism. This fundamental discovery, reinforced by the findings of Lynn Margulis and other microbiologists, was that the key factor that makes the biosphere work are the microbes. With this discovery in 1968 which was continued by Folsome and other researchers during the 1970s, a vital element in the science now called biospherics was revealed. The work that the Institute of Ecotechnics did during the next decade focussed on the elements of how to make such a created biospheric system. One approach taken was to consider the biological/atmospheric component of man-made biospheres as an apparatus. Biospheric systems increase free energy inside a materially closed apparatus if you have a throughput of energy from outside, as do both the Earth and Biosphere 2. The Second Law of Thermody-

namics is not violated because biospheres are not closed systems. Conversely, for analytic purposes, the technospheric unit as a behavioral region, is treated as an "engine" or, fundamentally, an entropy-producing component. If the increase in free energy of the life systems is greater than the entropy of the supporting technics, then basically we would have a biosphere that can continue indefinitely in harmony with its technosphere.

I formulated as a theoretical basis for biospheric systems the following three laws of biospherics, which can be tested in Biosphere 2 and subsequent biospheric systems. They are:

1. The energy passing through the system increases the free energy in the system relative to the entropy during the passage of time.

2. The system uses this free energy to increase its potential to extract a higher rate of free energy during the passage of time out of the incoming energy flux by a) increasing its mass by converting inorganic matter into organic matter, and b) by converting the inorganic matter

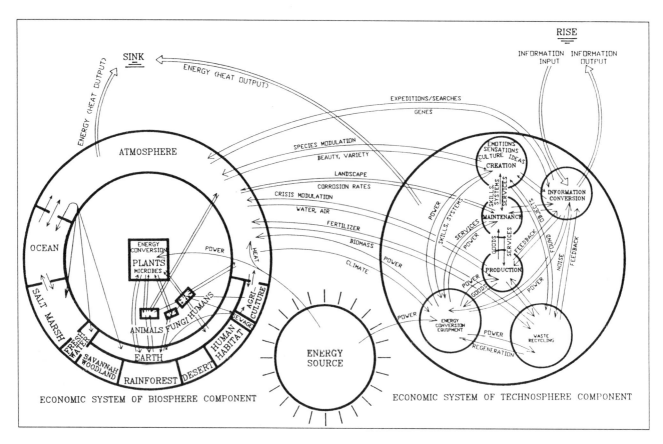

Figure 1. Biosphere/Technosphere Model for Biosphere 2. (Copyright 1986 by Space Biospheres Ventures)

into systems capable of storing more free energy.

3. Information passing through the system obeys the same laws of increasing free energy of the system during the passage of time, and of increasing the system's potential to extract a higher rate of free energy out of the incoming information flow during the passage of time.

Systems which do not obey these laws are inorganic systems or technical systems or failing biospheric systems; that is, the entropy is increased relative to the free energy during the passage of time upon the introduction of a flux of energy through the system.

Perhaps it is fortuitous, but more probably synchronistic, that the information revolution was occurring at the same time that Biosphere 2 was designed. Certainly space exploration, global studies and the creation of a complex system like Biosphere 2 is almost inconceivable without the integration of global electronic communications and the varied powers accessible through computers and computer networks. Besides allowing for an energy sink outside Biosphere 2, we looked for not only an information rise in an artificial biosphere, but an information rise outside Biosphere 2 by making a network of information between researchers inside Biosphere 2 and those in Biosphere 1 (as we have termed the biosphere of Earth). The information sink or noise will be converted to waste heat (erased programs and data) and thus join the energy sink. Information rise produced by converting data and information to knowledge and by evolution of ecological organization in the life systems is another addition to the free energy component of the system (Figure 1).

When the Space Biospheres Ventures team in 1984-5 translated these approaches and the experience gained by the Institute of Ecotechnics and other consultants to the project into a model of Biosphere 2, we came up, via several iterations, with a plan for a seven biomic area, 3.15 acre airtight structure, with an volume of about seven million cubic feet (Figures 2, 3, and Tables 1, 2). To make the necessary calculations, SBV worked out a 12 level hierarchy scheme of ecology. This includes the levels of microbes, multicellular species, populations, food web niche guilds, functional systems, patches, phases, communities, ecosystems, bioregions, biomes and finally, the biosphere. For practical design of artificial biospheres it is especially important how you use the functional ecosystems landscaped or bio-regioned by biomes.

Each of the levels has a different spatial and temporal scale. For example, we know that biospheres can operate on a billion year scale. Biomes operate on a scale that ranges from tens to hundreds of million years. Landscapes are component parts of biomes, and the time/space scaling descends progressively, down to the microbes at the bottom level which can have doubling times as

Figure 2. Biosphere 2 longitudinal section showing wilderness biomes, right to left: tropical rainforest, savannah (at top of rock cliffs), marine, marsh, and desert. Section measures 539 feet. (Copyright 1986 by Space Biospheres Ventures)

low as five minutes. In addition to this variety of time scales, there are differing spatial scales to keep in consideration. Biosphere 2 was designed for a minimum hundred year life span. Seven million cubic foot volume is the space scale that (according to our calculations based on mesocosm and Test Module work) is required to set up a situation which all the type phenomena associated with the biosphere might be produced and sustained.

Of course when we start out creating a biosphere today, it is quite different than the origins of our planetary biosphere some 3.8 billion years ago. There are many biologists who contend that the biosphere and the biotic cycle came before specific life forms, that perhaps clay molecules were recycling and building free energy as much as 200 million years before the origin of life. Using the clay as a template, the organic molecules could begin their reproductive processes.

In designing Biosphere 2, the SBV team had to include humankind and technics, besides the naturally occurring biomes. Thus, the work with Biosphere 2 can address the serious issues facing humanity in its relations with the Earth's biosphere, as well as providing valuable baseline data on how such systems operate as a preliminary to their design and creation for space habitation.

SBV had the challenge of developing two forms of intelligence to operate this biosphere/technosphere system. One was using the artificial form of intelligence. For this SBV's Computer Team, headed by Norberto Alvarez-Romo, developed a five level system hierarchy. The five functional levels identified are: 1) point sensing and activation,

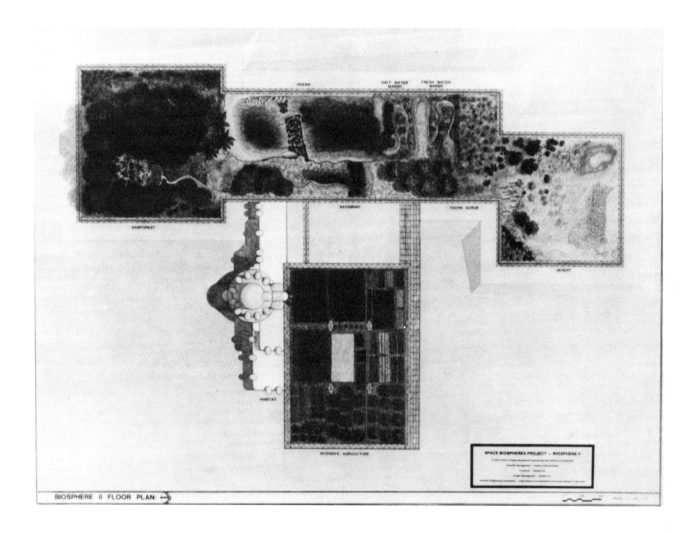

Figure 3. Biosphere 2 Floor Plan, showing wilderness biomes at top, human habitat and intensive agriculture biomes at bottom. (Copyright 1986 by Space Biospheres Ventures)

2) local data acquisition and control, 3) system supervisory monitoring and control, 4) global monitoring and historical archive, and 5) telecommunications.

In addition, to prepare for Biosphere 2 operations, unique computer software has been developed for the following areas:

a. Atmospheric Carbon Dioxide Modeling and Real-Time Monitoring of Bioregenerative Life Support Systems. This model has been used to simulate and predict carbon dioxide levels for experimentation (including manned closures) in the SBV Test Module and to assist design and engineering calculations for Biosphere 2.

b. Thermodynamic Modeling, Simulation and Real-Time Control in Bioregenerative Life Support Systems. Biospheric systems are open to information and energy exchange with the outer environment. Energy inputs for photosynthesis, electrical power, communications, and heating are required. There is a need to offset external fluctuations and dispose of waste heat. Internal relative humidity and energy efficiency must also be managed. Machinery for heating and cooling air tends to be complex enough to typically require a dedicated human staff for operations, monitoring and maintenance. SBV has developed BIOSYS (a thermodynamic simulation model for closed bioregenerative life support systems) and Real-Time Expert System Applications to reduce the labor required for such functions by integrating simulation and control of internal environmental parameters with computer-driven programs.

c. Global Monitoring of Closed Bioregenerative Life Support Systems. SBV has developed automated monitoring and diagnosis of overall life system status. In addition, an historical archive database combines diverse data sets: environmental, analytical and biological. SBV has created a Bioaccessions Database to inventory and keep a history of all biological introductions into Biosphere 2. Atmospheric and water quality must be monitored not only for real time levels but also for trends and expected behavior. The Global Monitor and Advisor also serves as a repository in which models of bioregenerative and technical processes can be tested in real-time simulations.

BIOSPHERE 2 AREAS	square feet	square meters	acres	hectare
Glass Surface	170,000	15,794	3.90	1.58
Footprints				
Intensive Agriculture	24,020	2,232	.55	.22
Habitat	11,592	1,077	.27	.11
Rainforest	20,449	1,900	.47	.19
Savannah/ocean	27,500	2,555	.63	.26
Desert	14,641	1,360	.34	.14
West Lung (airtight portion)	19,607	1,822	.45	.18
South Lung (airtight portion)	19,607	1,822	.45	.18
TOTAL Airtight Footprint	137,416	12,766	3.15	1.28
Energy Center	30,000	2,787	.69	.28
West Lung (weathercover dome)	25,447	2,364	.58	.24
South Lung (weathercover dome)	25,447	2,364	.58	.24
Ocean Water Surface Area	7,345	682	.17	.07
Marsh Surface Area	4,303	400	.10	.04

BIOSPHERE 2 VOLUMES	cubic feet	cubic meters
Intensive Agriculture	1,336,012	37,832
Habitat	377,055	10,677
Rainforest	1,225,053	34,690
Savannah/Marsh/Marine	1,718,672	48,668
Desert	778,399	22,042
Lungs (at Maximum)	1,770,546	50,137
TOTAL	7,205,737	204,045

Table 1. Areas and Volumes of Biosphere 2.

d. Nutrition Diet Planning and Crop Production Scheduling. Within a closed ecological system, space and facilities for providing adequate nutrients for humans is limited. The aim of the system is to provide a schedule for planting crops so that each harvest yields an appropriate quantity and combination of foods for optimal daily nutrition.

In addition to this computer/technical system, we trained the biospherian crew in ecological/naturalist observation. This is colorfully phrased by E.O. Wilson of Harvard as "the naturalist trance". In this particular state of attention, the scientific observer can begin to take in the totality of the life events occurring around him, and receive insights into its mechanisms and patterns.

SBV works with this parallel structure during training of the Biosphere 2 crew so that they can work as naturalist observers as well as with the artificial intelligence system. The control system is designed where there could be a human intervention at each stage of the computer hierarchy. The analytic/computer system can sound alarms and intervene if it discerns dangerous trends or conditions before the human observers do. The data from Biosphere 2 will be networked in real-time with scientific institutions which consult to SBV and to others in related fields. This will be important for research purposes and to also help detect incipient problems. It is inevitable, of course, that both the human being and the computing systems can give the wrong data/or reach false conclusions. Building this "binocular vision" of naturalist observation and artificial intelligence into the operation of Biosphere 2 increases the likelihood that at least one eye, hopefully, is working properly to monitor and manage the system, or if both should fail, that recovery will be quicker.

We had to revise our approach to the entire technosphere as we encounter it in the world today because the technosphere inherited from the Industrial Revolution, which began when the world had less than a billion people, is polluting the entire life environment in Biosphere 1. In Biosphere 1, though, the buffers or surge tanks or reserve capacities are so great that the time it takes for these impacts to reach the politically effective majority of human beings is quite long after they commence. In Biosphere 2 cycling times are faster and buffers much smaller. We cannot afford to have environmentally damaging technics in the system at all. (Neither can Biosphere I for much longer.) So SBV had to do quite a bit of work developing a technosphere which could give backup support to Biosphere 2 without polluting it. The challenge was to develop technical systems that would be adequate to the 21st century space exploration that Dr. Paine has described, while maintaining and helping better manage a healthy planet Earth biosphere.

The design of the cross-section of Biosphere 2 connecting Earth and Mars was the first logo of Space Biospheres Venture. *Logos,* I understand in its root meaning, denotes the structure of effective reason, and the structure of our effective reason was that biospheres constitute an essential component of living permanently in space. This is what our destiny, our adventure and our future in space requires.

Why did we pick Mars? Our conclusion came out of many of the same factors that persuade many other space scientists and thinkers who see the enormous potentiality of Mars. We discussed options with a number of people, the astrogeologists, astronauts, biologists and many other people with quite a profound interest in space. We could, of course, have started out and said let's first make a prototype for microgravity, let's use an opaque system and make, not a full biosphere, but a small ecosphere, with only an agricultural/atmosphere-regenerating life system. But we reasoned (and had many friends, astronauts, cosmonauts, astrogeologists, concur) that the objective, at once so doable that it would catch the imagination of hu-

	HIGH		LOW	
	Celsius	Fahr	Celsius	Fahr
Rainforest	35	95	13	55
Savannah	38	100	13	55
Desert	43	110	2	35
Intensive Agriculture	30	85	13	55

Table 2. Biosphere 2 Temperature Ranges

manity sufficiently to unlock the necessary resources, would be to set as a goal the settlement of Mars. SBV began the Biosphere 2 project with the idea that it would be directly sunlight-driven, modeled such that we would get valuable ecological knowledge for the Earth as well as developing further our conception of a well-developed Mars habitation base.

By 1987, Space Biospheres Ventures felt that we should have even broader scale discussions and interchange with the international community interested in closed ecological systems. We invited space life scientists from NASA, ESA, the Soviet space program, leading ecologists and scientists like Howard T. Odum, Ramon Margalef, Walter Orr Roberts, pioneers in the field like Clair Folsome, Ganna Meleshka (Institute of Biomedical Problems, Moscow) and Josef Gitelson (Bios-3 Project, Institute of Biophysics, Krasnoyarsk) to participate in an international workshop on closed ecological systems at the Royal Society in London. In September, 1989, the second international closed sys-

tems workshop was held in Krasnoyarsk, Siberia, co-sponsored by the Institute of Biophysics (IBP), Institute of Ecotechnics and SBV. Gitelson, the director of IBP had told us this meeting would coincide with a major new step in *glasnost* that was going to be opening up Krasnoyarsk to travel from the outside and to the international scientific community. Indeed there was an almost incredible openness in the workshop, remarkable not only because so many of the Soviet scientists had studied English so that it could be the official language of the meeting, but because a free and full examination of the Bios-3 closed system facility where the most advanced Soviet closed system work has been conducted was allowed. We had previously opened the Biosphere 2 site to our Russian colleagues.

At the Krasnoyarsk meeting, the participants issued a resolution recommending that the name "Biospherics" be used for the scientific discipline which studies, creates and manages closed ecological systems. These include CELSS-type sys-

Figure 4. The SBV Biospheric Research and Development Center at the Biosphere 2 Project.

tems, "ecosphere" closed objects that have one ecosystem, and biospheric closed objects that have two or more ecosystems with ecotone interaction, both small man-made (Biosphere 2 and its successors on Earth and elsewhere) and large natural biospheres (the Earth's).

In 1986 SBV constructed the Biosphere 2 Test Module, designed for two purposes. One was to check out the sealing and structural engineering planned for Biosphere 2. The second was to be a testbed for research in the operation of closed ecological systems. We calculated that in its volume, and using mainly sunlight levels of energy, we could design life and technical systems to support one human being within an ecosystem. There has been closed life system research in the Test Module since the end of 1986, included three, five and 21 day human closure experiments.

One of the major problems in building Biosphere 2 was — and there were a number of problems! — there was no research facility that we could subcontract to conduct a lot of the preliminary investigations. So we had to build an entire research facility called the Biospheric Research and Design Center (BRDC) at the project site (Figure 4). This facility includes computer laboratory, plant tissue culture and analytical chemistry laboratories, insectary, and plant quarantine facilities in addition to the Test Module prototype and agricultural/aquaculture greenhouses to develop cropping systems and techniques (Figures 5, 6). The work at BRDC accelerated the design and logistics of the Biosphere 2 project. In addition, research was conducted by the many research scientists, engineers and institutions consulting to Space Biospheres Ventures. Since there were not any existing research institutes focussed specifically on biospherics and the creation of closed ecological systems, SBV faced at least some of the problems that NASA faced in selecting an astronaut corps at the

Figure 5. Biosphere 2 prototype agricultural system.

beginning of the space age. The Institute of Ecotechnics' contributed in this regard, as a consultant to SBV, by recommending people and institutions whose work was known by the Institute of Ecotechnics through its conferences and field projects consultancies over a number of years. At present, SBV is moving ahead quite rapidly and simultaneously in research and development, systems design and architecture, construction and quality control, and biospheric training programs. Such biospheric systems and the mastery of clos-

ed life systems are needed to open the road along with astronautics to widen the horizons of life in space.

The Biosphere 2 time scale calls for completion and closure for the first two year experiment in the fall of 1990 (Figures 7, 8). At that time, we will begin work on opaque systems research using the biospherics expertise gained in the Test Module and Biosphere 2. These opaque life systems will be oriented towards space station, lunar base and extended planetary mission use; that is, towards

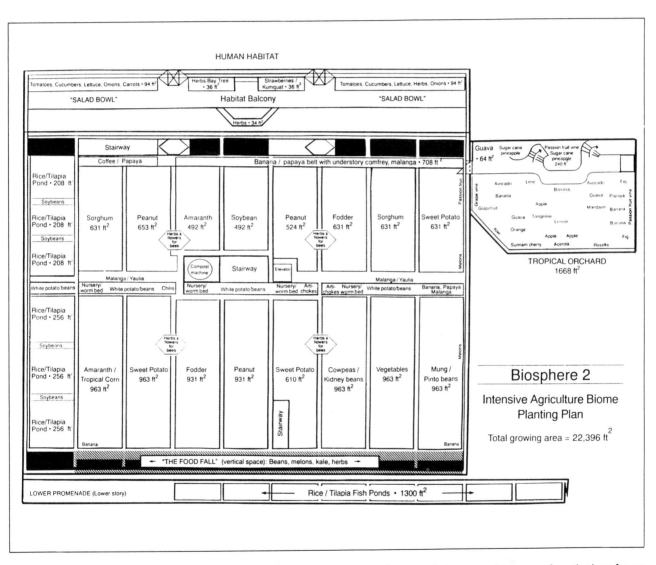

Figure 6. Biosphere 2 Intensive Agriculture Planting Plan. Plant growing area is increased by use of vertical surfaces which border on technospheric areas not requiring sunlight, and use of the portion of sloped lower story which receives full sunlight. A two-year plant cropping scheme designates harvest and planting schedules, rotation of crops, and recycling activities over the initial manned closure experiment to acheive human and domestic animal nutrient requirements and maintain soil fertility. Planting plans vary according to season of the year. (Copyright 1989 by Space Biospheres Ventures)

space applications that are relatively near-term. We would then look at the testing and deployment of such life systems in microgravity. Space Biospheres Ventures has a small investment in the External Tanks Corporation, and is negotiating joint ventures with several companies, American and foreign, who are working in these space application fields. Space Biospheres Ventures is also a founding corporate member of the International Space University, and has helped organize workshops on closed ecological systems at the last two conferences of the Space Studies Institute in Princeton. When we commenced the design and construction phase of SBV in December 1984 SBV targeted the early 1990's as when we wanted to be ready with closed systems because we considered the mid-1990's would see the operation of the microgravity space stations. By 1992 we think SBV will be ready to produce ecosystem modules for microgravity and initial lunar deployment.

The financing of Biosphere 2 may be of interest to NASA and others here from universities and private corporations. All space-related work to date has had tremendous commercial spinoff and so SBV decided to finance Biosphere 2 by venture capital. SBV anticipates the spinoff from commercial applications of biospherics on air, soil and water pollution control, environmental control, software systems for monitoring and management of complex systems in addition to the education and training programs that will come out of Biosphere 2 will provide very good returns on investment.

On behalf of SBV, I want to welcome all of you to the Biosphere 2 Project. We hope that good relationships and interchange occur during the workshop which can continue into the future. We

Figure 7. Biosphere 2 under construction, exterior view.

know that the action of everyone here is extremely important for the space and Earth objectives that are becoming possible. At this workshop are represented university, governmental and private industry research and application groups. We think that the cooperation of these different kinds of human institutions is going to be just as necessary as the working together of different peoples or scientific fields. An extraordinary range of efforts by individuals and institutions is needed to make this transition into our solar system home, humanity's first great step to the stellar world.

Figure 8. Biosphere 2 wilderness biomes under construction, Interior view. Savannah at right; marsh biome upper left, marine biome lower left.

Biosphere 2 Test Module Experimentation Program

Abigail Alling, Linda S. Leigh, Taber MacCallum and Norberto Alvarez-Romo*
Space Biospheres Ventures
Oracle, Arizona

The scale of closed ecological system experiments to date has ranged from studies with 100 ml systems to the largest existing system — Space Biospheres Ventures' Biosphere 2 Test Module, a variable volume facility of some 480 cubic meters. The science of materially closed ecological systems started in 1968 with Prof. Clair Folsome's ecosphere work at the University of Hawaii. Folsome began by sealing aquatic microbial assemblages in 100 ml to 5 liter flasks and exposing them to indirect sunlight. These ecospheres have remained indefinitely viable; the oldest are now over 20 years old, demonstrating that closed ecological systems can persist over time with an input of energy. The CELSS research program pioneered by the National Aeronautics and Space Administration includes studies of biomass production with higher plants and other aspects of bioregenerative life support. The largest current testbed is the Breadboard project at Kennedy Space Center [see William Knott's presentation] where studies are conducted within a closed 3.5 meter by 7.5 meter cylindrical, steel biomass production chamber. The Institute of Biophysics at Krasnoyarsk, Siberia has also experimented with a 300 cubic meter steel structure (Bios-3) with closures of up to two to three people for six months. The aim of Bios-3 was to establish a near complete air and water regeneration with considerable food production.

It is in this context, that the Biosphere 2 Test Module research at Space Biospheres Ventures is significant. Over the past four years, progressive research has been conducted within the Biospheric Research and Development Center to design and test a total system approach to closed ecological systems research. In the Biosphere 2 Test Module the first experiments to utilize completely biological methods of air, water and waste regeneration and food production were conducted.

THE FACILITY

The Biosphere 2 Test Module is the largest closed ecological research facility ever built with a sealed variable volume of some 480 cubic meters and a unique steel and glass skin which allows an average of 65% of ambient Photosynthetic Active Radiation to penetrate into the system (Figure 1). It was designed to test both ecological and engineering systems developed for Biosphere 2. During the early phases of research, the physical structure itself was under investigation. The nature of closed system research necessitated that the Test Module be sealed and that SBV develop a method to determine leak rates.

The sealing techniques utilized by SBV underwent considerable development. The first system employed, which utilized butyl rubber sealants on

---

* Prepared by Abigail Alling, Director of Marine Ecological Systems, Space Biospheres Ventures; Linda S. Leigh, Director of Terrestrial Ecological Systems, Space Biospheres Ventures; Taber MacCallum, Analytical Systems Manager, Space Biospheres Ventures; and Norberto Alvarez-Romo, Director of Cybernetics Systems, Space Biospheres Ventures. Presented by Abigail Alling.

single-paned glass joints with the steel spaceframe members, provided as tightly sealed a structure as any which had previously been utilized in the field. It was judged inadequate to the task of providing the sealing for the 3.15 acre Biosphere 2 structure, with its over 20 miles of glass and steel seals. A second system, patented by SBV, was developed which increased the sealing efficiency considerably. When last measured in March/April 1989 the Test Module had a leak rate of about 24 percent per year (a turnover of air in a little over four years) which when translated to the larger volume per structure ratio of Biosphere 2 gives a projected leakrate of about three percent per year.

Similarly, SBV moved from a single-paned glass system used on the first Test Module roof to a double laminated glass system after two of the original single panes cracked about a year after installation, probably as a result of hairline factory fractures. Initially, it was also planned to install a louver system on Biosphere 2. This was tested on the first Test Module roof structure but was eliminated on the present structure because of the reduction in incident light it caused.

SBV also had to develop a method of managing the effects an internal temperature and external barometric pressure change could cause in a fixed, sealed, glass structure. This problem was solved with a design called a "lung", a variable volume system joined to the module by an air duct. With increased temperature or decreased barometric pressure in the Test Module compared to the outside environment, the variable chamber expands; with a decrease in temperature or a increase in pressure, the chamber contracts. The lung structure provides an effective means to prevent the possibility that the Test Module would implode or explode when subjected to these forces. Further, the reservoir of air provided an increased buffering; adding approximately 20-40% to the total atmospheric volume. Further, the weight of the pan on the lung structure insured a positive displacement from inside the closed system to the outside.

LIFE SYSTEM RESEARCH

Following the structural research during 1986, the next two years focused on studies of higher plants

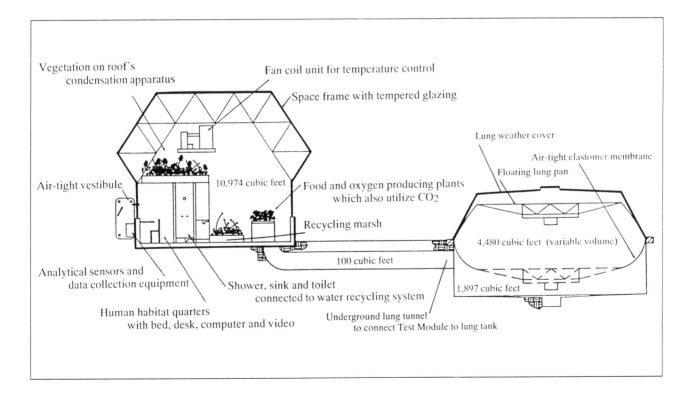

Figure 1. Biosphere 2 Test Module System Schematic. Surface area for plant growth area includes mezzanine platforms and areas within hollows of the space frame. (Copyright 1987, Space Biospheres Ventures.)

and soils and their interaction with the atmosphere, light, temperature and community structure. On December 31, 1986, the first of a series of ecological experiments commenced which lasted up to three months in duration. During these closures questions included:

1. Would plant species reproduce in a high humidity environment?

2. Would plants and soil microbial filters manage to remove trace gases from the atmosphere?

3. What effects would reduced ultraviolet light have on the behavior and navigation of foraging bees?

4. Would all the functions of microbes within the soil ecology be present?

This series of experiments provided information for doing detailed modeling of the basic parameters of closed ecological systems.

In September, 1988, Space Biospheres Ventures took a further step which was to include one human (John Allen) in the closed ecological system for 72 hours (Figure 2). The system was 100% closed with respect to water, food and air. All waste materials were recycled in the Test Module using a marsh recycling system developed in consultation with Bill Wolverton of NASA Stennis Space Center. When John Allen exited the Test Module after 72 hours, looking healthy and relaxed, and our sensors showed no buildup of potentially toxic trace gases, we knew we were on the way to establishing systems which not only support human life, but which create a habitat conducive to human life. We stress this point because an important key to living for extended periods off this planet will be the development of life systems which can provide humans with not only all of the physiological requirements of life support, but ones which are satisfying to live in as the terrestrial ecology that we are adapted to. The ecology within the Test Module system was such a design.

Drawing on the research of our Russian colleagues, we knew that algae and higher plants

Figure 2. Biosphere 2 Test Module.

Biological Life Support Systems

were able to regenerate the oxygen required by human life while removing respired carbon dioxide. Never before, however, had waste materials been treated within the closed system and air completely recycled with biological methods. Following the design of Biosphere 2, we used higher plants and soils to recycle the atmosphere. For the first time, soils were introduced into closed system ecology and designed by SBV to be a primary bioregenerative system using soil bed reactor technology patented by SBV. Not only was carbon dioxide managed using this system, but trace organic gases and potential toxic gases were kept within acceptable concentrations for human and plant life.

Table 1 shows the type of range of trace gases found during our Test Module closures involving a human occupant. These gases were identified using a gas chromatograph mass spectrometer and a gas chromatograph flame ionization detector. In all of our closures none of these gases reached levels considered toxic to human life as defined by OSHA and the American Conference of Governmental Industrial Hygienists. However, monitoring gases with continuous sensors has been a significant challenge. In the first experiment sensor drift and noise made the continuous monitoring unreliable; we had to rely on recalibration of the entire system to locate the actual concentrations. These recalibrations were always far under concentration levels considered to be of concern. In the second "Human in Closed Ecological System Experiment", a five day closure with Gaie Alling, SBV changed the entire system using sensors internal to the Test Module, but these were evaluated as still not at the level required for this type of research and especially not for Biosphere 2. The third Human in Closed Ecological System Experiment began on October 26, 1989 for a one week material closure prior to the three week human closure (Linda Leigh) from November 2-23. A week closure followed her exit from the facility. For this experiment SBV developed an analytical system which is achieving to date a continuous and reliable record of 11 critical gases: CH_4, total non-methane hydrocarbons, NO_x, O_3, NO, CO_2, O_2, H_2S, SO_2, NO_2, and NH_3.

Data from this 21 day human closure experi-

Table 1: Trace organic gases identified by three methods in the SBV Human in Closed Ecological System Experiment, September 10 - 30, 1988.

Probable Origin: a = Technogenic • b = Biogenic • c = a + b

A. Identified by Gas Chromatograph/Mass Spectrometer

Compound	Number of Isomers Found	Probable Origin
Alkyl Substituted Cyclopentane	1	c
2-butanone	1	c
Carbon Disulfide	1	b
Cyclohexane	1	c
Decahydronaphthalene (decalin)	1	a
Decamethylcyclopentasiloxane	1	a
Decane	1	c
Dimethylbenzene	2	a
Dimethylcyclohexane	3	c
Dimethylcyclopentane	4	b
Dimethylhexane	2	c
Dimethyloctadienol Acetate	2	b
Dimethyloctane	2	c
Dimethyloctatrine	1	b
Dimethylpentane	1	b
Ethylmethylcyclopentane	1	c
Ethylbenzene	1	c
Ethylcyclohexane	1	c
Heptane	1	c
Hexamethylcyclotrisiloxane	1	a
Hexane	1	c
Isopropyl Substituted Cyclopentane	1	b
Methyl (methylethenyl) Cyclohexane	1	b
Methylbenzene	1	a
Methylbicyclohexene	1	b
Methylcyclohexane	1	c
Methylcyclohexene	1	c
Methylcyclopentane	1	c
Methylheptane	1	a
Methylhexane	2	c
Octamethylcyclotetrasiloxane	1	a
Substituted Cyclohexane	3	b
Substituted Cyclohexene	1	b
Tetrachloroethene	1	a
Tetrahydrofuran	1	a
1,1,1 Trichloroethane	1	a
Trichloromethane	1	a
Trimethylbicycloheptene	1	b
Trimethylcyclohexane	2	c
Trimethylcyclopentane	3	b
Trimethylpentane	1	c
Trimethylsilanol	1	a

B: Identified by Gas Chromatograph/Flame Ionizer Detector

Ethane	1	c
Ethylene	1	c
Methane	1	c
Propane	1	a

C: Monitored with continuous sensors

Ammonia	n/a	b
Carbon Monoxide	n/a	b
Formaldehyde	n/a	a
Hydrogen Sulfide	n/a	b
Nitrogen Dioxide	n/a	b
Ozone	not detectable	
Sulfur Dioxide	n/a	b

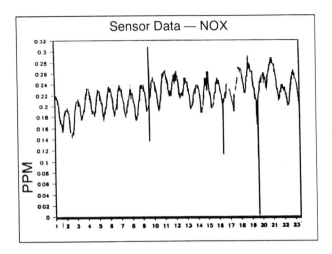

Figure 3. NOx levels during Test Module human closure November 2-23, 1989.

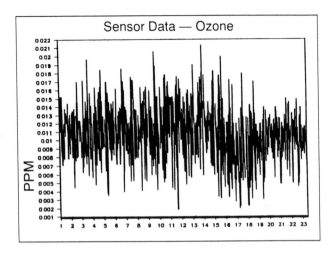

Figure 4. Ozone levels during Test Module human closure November 2-23, 1989.

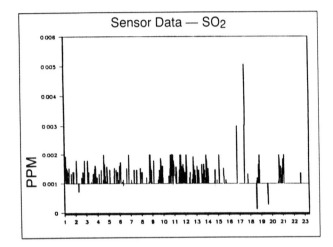

Figure 5. Sulfur dioxide levels during Test Module human closure November 2-23, 1989.

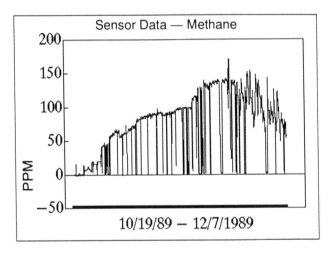

Figure 6. Methane levels during Test Module human closure November 2-23, 1989 including unmanned closure phases pre- and post-human habitation.

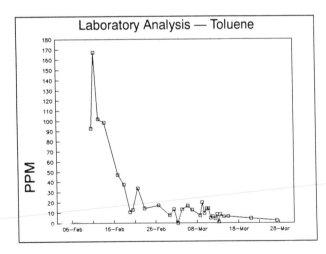

Figure 7. Toluene levels during Test Module human closure March 8-13, 1989 including unmanned closure phases pre- and post-human habitation.

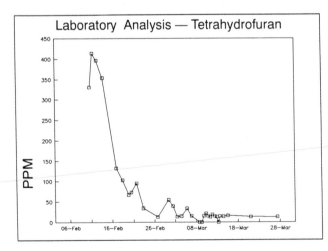

Figure 8. Tetrahydrofuran levels during Test Module human closure March 8-13, 1989 including unmanned closure phases pre- and post-human habitation.

ment on trace gas levels is illustrative of the low levels maintained in all our experiments. Figure 3 shows nitrogen oxides (NOx) concentrations which ranged from 0.15 to about 3 ppm. Cautionary eight hour levels are considered to begin above 30 ppm.

Figure 4 shows ozone levels which show highs of 0.021 ppm. Cautionary levels begin at 0.1 and danger levels at 0.3 ppm.

Figure 5 is a graph of sulfur dioxide where levels stayed below. 0.005 ppm — well below the alert levels of 2-5 ppm.

Figure 6 is methane. The slight rise shown to about 150 ppm (still far below those of concern) during the human closure corresponds to other experiments conducted at SBV and at the Environmental Research Laboratory, University of Arizona. For methane, data suggests the hypothesis that it takes some time before the methane-metabolizing microbes build up their populations to bring down atmospheric concentrations. It then forms a classic negative feedback loop.

Figures 7, 8 and 9 are typical of the data we obtain for technogenic gases in the Test Module. Figure 7 shows toluene, which often is found in the outgassing from paints. Figure 8 is tetrahydrofuran, a solvent, often implicated in the "sick building syndrome", and which is released from glues used in such things as carpets and plywood. Figure 9 shows ethyl benzene, a solvent used in resins, probably an outgassing from particle board and plywood. All three gases show, in these graphs from our March 1989 experiment, an initial rise after closure, following the flushing of the Test Module air. Then they are quickly brought down to extremely low levels by the action of soil bed reactors and other biological metabolizers.

The major subsystems of the Test Module designed for human closure experiments include the following:

Human Habitat Living Quarters

In addition to providing basic accommodations, the Test Module human habitat was designed to allow the human resident to observe and participate in manned closure experiments as a researcher (Figure 10). Human living quarters are comparable to a small efficiency apartment plus a compact workstation. Within an area of 100 square feet, the habitat includes:

1. a small kitchen (microwave oven, electric induction coil heat plate, electric water heating urn, small refrigerator, sink with hot and cold potable water, food weighing and preparation counters, and utensils);

2. a "Murphy" bed which folds up into a self contained wall cabinent when not in use;

3. a water-conserving toilet, and shower which uses only 0.9 liters/minute of water;

4. a workstation area with computer, desk, and bookshelves;

5. telecommunications systems — telephone, video link for audio/visual teleconferencing, computer links to internal SBV networks (to access the analytic monitoring system and databases) and to external telecommunications networks;

6. basic human physiological monitoring apparatus which vary according to the experiment.

Analytic System

The analytic requirements include a continuous monitor of the eleven trace gases with continuous atmospheric sensors. In addition other trace organics in the air were tested once a day using a gas

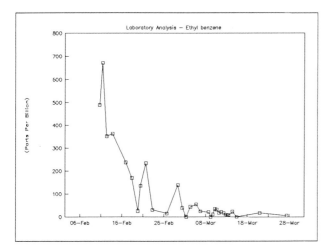

Figure 9. Ethyl benzene levels during the Test Module human closure March 8-13, 1989 including unmanned closure phases pre- and post-human habitation.

chromatograph and ion chromatograph system. Testing of the potable, recycle and irrigation water quality is done once a day as well.

Life Systems

The ratio of carbon dioxide consumed (photosynthesis) to carbon dioxide produced (respiration) must be greater than one before introduction of the human so that the system can compensate for the 850-1100 grams of carbon dioxide (depending on body weight, diet, level of activity) exhaled by a person each day; to provide high quality potable water through condensation of the evapotranspired water of the plants; and to provide all a person's nutritional needs. The Test Module life system designs for human closure have included the following (Figures 11 and 12):

1. <u>Plants</u>. Plant species were chosen with a high growth rate, high photosynthetic rates and selected at a young growth phase to maximize the amount of carbon dioxide which could be utilized by each plant.

a. Included in this design were the following sub-systems:

1) savannah mezzanine area with C_4 grasses, adapted to high temperatures and light levels
2) intensive agricultural plants such as sweet potatoes, sugar cane and peanuts which have very high growth rates, as well as a range of other vegetable, bean, salad and grain crops,
3) a "ginger belt" which includes the fast growing zingerberacae order plants, such as banana, ginger and canna, and
4) marsh recycling system with water hyacinth as the dominant species.

b. A focus of some of our experiments has been to examine the production and activity of methane within the Test Module. The November

Figure 10. Biosphere 2 Test Module Human Habitat during the first "Human in Ecosystem" experiment, September 1988. John Allen prepares a meal in the Test Module habitat kitchen.

1989 closure included a 2.6 square meter marsh system and a 0.65 square meter rice paddy with Tilapia fish. Methane dynamics are of great concern globally as methane is a component of the greenhouse effect and its quantitative outputs from known sources like marshes and rice paddies is poorly known.

c. A bioaccessions list, computer linked, inventoried all the plant species introduced. Biomass determinations of soil and foliage were made at closure and upon completion of the experiment.

2. <u>Soils</u>. To decrease the amount of soil respi-

ration, soils were composed with low organic carbon and a high nutrient mixture of pumice, natural soil, and bat guano.

Monitoring System

The computer monitoring system (termed the "nerve system") design has access to varied sensors which relay information in a five-level structure to a command center located in the SBV Mission Control building. The five functional levels are 1) point sensing and activation, 2) local data acquisition and control, 3) system supervisory monitoring

Figure 11. Biosphere 2 Test Module interior view, ground story. Marsh waste recycling system at lower left; part of "ginger belt" at lower right. SBV Researcher Linda Leigh stands next to intensive agriculture system planted in the soil bed reactor (white planting box). Savannah mezzanine section with C_4 grasses is located above.

control 4) global monitoring and historical archive and 5) telecommunications.

The G2 software controls and monitors continuous gases and checks the analytic sensor calibrations. G2 is also the program with which we are modeling carbon dioxide cycling. This program is dynamic and allows for a real time interaction to occur between our predictive model and the data as observed in the Test Module experiment. RTAD is the software designed for data acquisition and control.

Water Systems

The water recycling system consists of three subsystems: potable water, wastewater from the habitat, and plant irrigation water.

1. The waste recycling system provides complete recycling of all human wastes. With this system, no wastes are removed from the Test Module; the sewage, kitchen and domestic water is purified by the action of microbes and plants and then used to irrigate the plants in the Test Module. The system is designed to clean 5-15 gallons of effluent per day and during all three "Human in Closed Ecological System Experiments", the 2.6 square meter system effectively and without malodor cleaned the waste products using both anaerobic and aerobic processes.

Figures 13 and 14 present data from the operation of the waste recycling and irrigation water systems during the November 1989 experiment. They show levels of nitrates and phosphates in the aquatic waste processing system — after being held in the anaerobic holding tank where anaerobes start the process of regenerating the waste water, batch additions are made to the aerobic tank where the aquatic plants and their symbiotic microbes continue the process, bringing levels down so that the water can then be routed to the irrigation water system, while producing an abundant increase in plant biomass. Data from the irrigation water samples show concentrations of nutrients rise after entry of the human into the system and

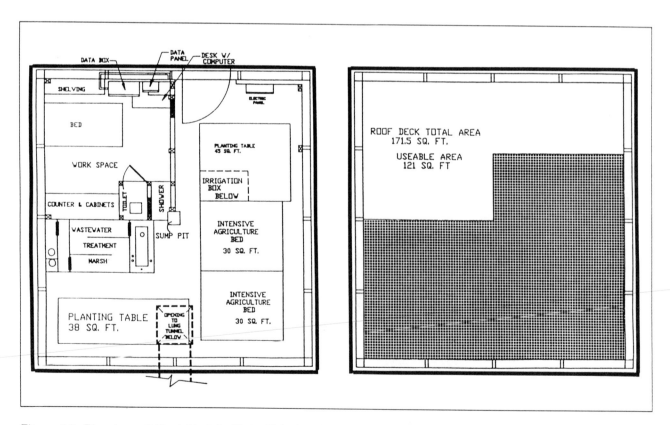

Figure 12. Biosphere 2 Test Module Floor Plan (excluding lung) during the "Human in Ecosystem" experiments conducted in September 1988, March 1989, and November 1989. (Copyright 1988, Space Biospheres Ventures.)

Biological Life Support Systems

periodic rises with batch additions from the waste recycling system followed by uptake by plants.

2. Potable water is distilled from the atmosphere by two dehumidifiers and sterilized with ultraviolet sterilizer systems. Potable water supplies all kitchen water as well as a 0.9 liter/minute shower.

3. Irrigation water includes all run-off water from life systems and some potable water. Water is held in a reservoir and pumped to the plants through computer controlled solenoid valves to various irrigation zones.

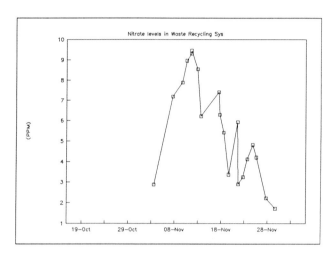

Figure 13. Nitrate levels in aquatic plant/microbial waste recycling system during Test Module human closure November 2-23, 1989 including unmanned closure phases post-human habitation.

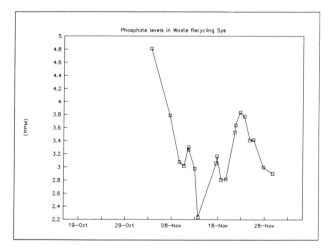

Figure 14. Phosphate levels in aquatic plant/microbial waste recycling system during Test Module human closure November 2-23, 1989 including unmanned closure phases post-human habitation.

In all these experiments, the inhabitants of the Test Module lived in material closure from the outside and depended on the ecology and technics within the Test Module to maintain the environment, recycle nutrients, the atmosphere, and water, and provide an esthetic and comfortable home. SBV has to date conducted over sixty days of human closure experiments in the Biosphere 2 Test Module.

SUMMARY

The Biosphere 2 Test Module is a facility which gives us the capability to do either short or long term closures; we have conducted five month closures with plants. We can also conduct detailed investigations of specific problems, such as trace gas purification by our bioregenerative systems by in-putting a fixed concentration of a gas and observing its uptake over time. In other Test Module experiments the concentration of one gas was changed to observe what effects this has on other gases present or the system. We are looking forward in the coming year after the completion of studies necessary for Biosphere 2 to use the Test Module for experiments related to near-term space applications, such as space station life support systems, technologies for extended planetary missions and initial lunar base requirements.

Until recently, humankind has not played a direct part in the management of the biosphere of the Earth, which we have termed Biosphere 1. Life itself has managed total system ecology in our global biosphere — particularly the microbes which play a great and frequently unappreciated role. Now humankind can and must participate in cooperating with the processes of the biosphere. The science of biospherics which encompasses the study of closed ecological systems provides an opening into the future in space as well as in our Earth's biosphere. Like the first steps that initiate all exploration, we have described these experiments in the Test Module — our first steps between Biosphere 1 and Biosphere 2.

Soil Bed Reactor Work of the Environmental Research Lab of the University of Arizona in Support of the Research and Development of Biosphere 2

Robert Frye, Ph.D.* and Carl N. Hodges (Director)
Environmental Research Laboratory, The University of Arizona
Tucson, Arizona, U.S.A.

INTRODUCTION

The Environmental Research Laboratory of the University of Arizona was engaged through the Planetary Design Corporation, on behalf of Space Biospheres Ventures, developers of Biosphere 2, to assist with certain aspects of the scientific design of the Biosphere.

The areas of our contribution range from assistance with general engineering questions to extensive supporting work for the Intensive Agricultural Biome and a major program on issues having to do with air purification and the ultimate composition of the atmosphere within Biosphere 2. The scientific work reported in this paper was conducted under the direction of Dr. Robert Frye and he has prepared the paper that I have the pleasure of presenting.

Carl N. Hodges, Director
Environmental Research Laboratory

SOIL BED AIR PURIFIER RESEARCH AT ERL

Research at the Environmental Research Laboratory of the University of Arizona (ERL) in support of Biosphere 2 has been both of a basic and applied nature. One aspect of the applied research has involved the use of biological "reactors" for the scrubbing of trace atmospheric organic contaminants. These "reactors" so named by Dr. Heinrich Bohn, University of Arizona, who did original work in this field, may be used in both open and closed environments. Our research has involved a quantitative examination of the efficiency of operation of Soil Bed Reactors (SBR) and the optimal operating conditions for contaminant removal.

The basic configuration of a SBR (Figure 1) is that air is moved through a living soil that supports a population of plants. Upon exposure to the soil, contaminants are either passively adsorbed onto the surface of soil particles, chemically transformed in the soil to usable compounds that are taken up by the plants or microbes, or the compounds are directly used by the microbes as a metabolic energy source and converted to CO_2 and water.

The number and type of compounds degradable by soils is large. Figure 2 is a compilation of compounds that are either known to be degraded in soils or are suspected to be degradable from in vitro studies. We have worked with only a subset of these compounds in our experiments: methane, ethane, ethylene, propane, carbon monoxide and nitrous oxide.

Our SBRs come in many sizes and shapes, some of our research has been conducted with large SBRs having a diameter of approximately one meter. Those shown in Figure 3 in a greenhouse at ERL have been used primarily to study methane removal and the effect of operating a SBR on plant growth and development. Our results to

---

* Discussion paper, scientific and technical work prepared by Dr. Robert Frye, Research Scientist, Environmental Research Laboratory, University of Arizona. Paper presented by Carl N. Hodges. This is ERL contribution #90-19R.

date indicate that a SBR has no impact on plant productivity or phenology. That means that functioning soils can be used for both intensive cropping (biomass production) and air purification — a most important result for their utilization in space life support systems.

The factors that would impact the functioning of a SBR are those that impact soil microbe physiology. Factors such as soil moisture levels, temperature, organic matter content, soil type and air flow through the SBR should be important in determining the efficiency of its operation. Our research has focused primarily on organic matter content, soil type, and air flow rate as easily manipulated variables. In addition, we have found that the history of the SBR's exposure to contaminants is important.

In our large format SBRs we conducted a long term study on the removal of methane from an incoming air stream. This experiment was undertaken to examine whether the operation of a SBR

declines with time. The graph in Figure 4 shows that with time a SBR becomes significantly more efficient at removal of methane. The three curves are fitted lines using the logistic population growth model. The implication of these results is that the efficiency of removal is dependent upon the population size of the microbial community in the soil and that upon exposure to a certain trace gas, that population increases over time. SBR #1 and #2 had different soil types which differed in organic matter content while SBR #3 had the same soil as in SBR #2 but only half the depth.

Another type of SBR we have used extensively at ERL is what we call our Aquaria SBRs (Figure 5). We have used these small systems to facilitate rapid acquisition of data which is not easily accomplished with the larger SBRs. These systems contain about 1.7 liters of soil in a container placed within a sealed 38 liter aquarium. The atmosphere within the aquarium is cycled through the soil with an aquarium pump. Flow rates of air through the

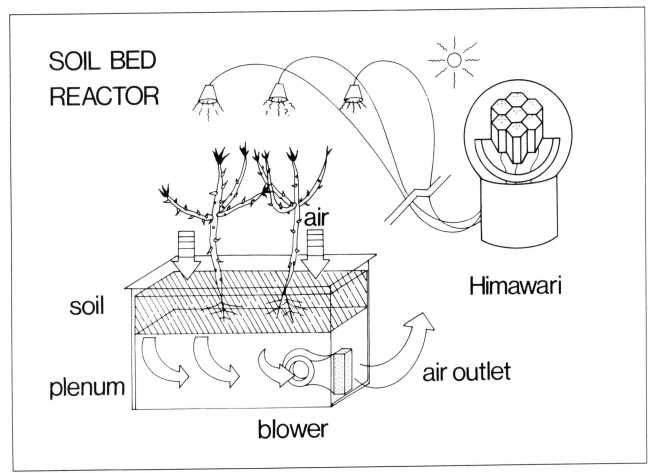

Figure 1. Schematic drawing of Soil Bed Reactor (SBR) for air purification.

Figure 2. Compounds known or suspected to be decomposed by soils or soil microorganisms.

Compound	Reference	Compound	Reference
Acetaldehyde	Fuller W.F. et al. 1983.	Ethylcyclohexane	Stirling, L.A. et al. 1977.
Acetic acid	Zavarzin, G.A. et al. 1977.	Ethylene	DeBont, J.A.M. 1976.
Acetoin	Bohn, H.L. 1977.	Flouro-4-nitrobenzoate (2-)	Horvath, R.S. 1972.
Acetylene	Smith, K.A. et al. 1973.	Flourobenzoate (o-)	Horvath, R.S. 1972.
Acrolein	Fuller W.F. et al. 1983.	Flouride	Bohn, H.L. 1977.
Alkyl benzene sulfonate	Horvath, R.S. et al. 1972.	Formaldehyde	Grundig, M.W. et al. 1987.
Aldehydes	Fuller W.F. et al. 1983.	Formate	Hou, C.T. 1980.
Ammonia	Hutton, W.E. et al. 1953.	Heptadecylcylcohexane	Beam, H.W. et al. 1974.
Anthracene	Dalton, H. et al. 1982.	Hexadecane	Beam, H.W. et al. 1974.
Benzene	Dalton, H. et al. 1982.	Hydrogen sulfide	Smith, K.A. et al. 1973.
Benzoate	Dalton, H. et al. 1982.	Hydrogen	Zavarzin, G.A. et al. 1977.
Bicyclohexyl	Higgins, I.J. et al. 1979.	Isoprene	Van Ginkel, C.G. et al 1987.
Bromomethane	Dalton, H. et al. 1982.	Isopropyl benzene	Higgins, I.J. et al. 1979.
But-2-ene	Higgins, I.J. et al. 1979.	Isopropylcyclohexane	Stirling, L.A. et al. 1977.
Butadiene (1,3-)	Van Ginkel, C.G. et al. 1987.	Isopropyltoluene (p-)	Horvath, R.S. 1972.
Butane	Hou, C.T. 1980.	Lactic acid	Bohn, H.L. 1972.
Butene (1-)	Dalton, H. et al. 1982.	Limonene	Dalton, H. et al. 1982.
Butene (cis-2-)	Dalton, H. et al. 1982.	Methane	Anthony, C. 1982.
Butene (trans-2-)	Dalton, H. et al. 1982.	Methanol	Dalton, H. et al. 1982.
Butylbenzene (n-)	Horvath, R.S. 1972.	Methyl mercaptans	Fuller W.F. et al. 1983.
Butyl-cylohexane (n-)	Horvath, R.S. 1972.Table 1.	Methyl sulfide	Smith, K.A. et al. 1973.
Butyric acid	Bohn, H.L. 1972.	Methylcatechol (3-)	Horvath, R.S. 1972.
Cadaverine	Bohn, H.L. 1977.	Methylcyclohexane	Stirling, L.A. et al. 1977.
Caprolactone	Stirling, L.A. et al. 1977.	Methylnaphthalene (1-)	Higgins, I.J. et al. 1979.
Carbon monoxide	Bartholomew, et al. 1982.	Methylnapphthalene (2-)	Higgins, I.J. et al. 1979.
Chlorobenzoate (m-)	Dalton, H. et al. 1982.	Napthalene	Dalton, H. et al. 1982.
Chlorofluoromethanes	Bohn, H.L. 1977.	Nitric oxide	Bohn, H.L. 1972.
Chloromethane	Dalton, H. et al. 1982.	Nitrous oxide	Goyke, N. et al. 1989.
Chlorophenol (m-)	Higgins, I.J. et al. 1979.	Ozone	Turner, N.C. 1973.
Chlorotoluene (m-)	Higgins, I.J. et al. 1979.	Octadecane	Perry, J.J. 1979.
Cinerone	Horvath, R.S. 1972.	Organophosphorus	Bohn, H.L. 1977.
Cresol (m-)	Higgins, I.J. et al. 1979.	Pentachlorophenol	Lagas, P. 1988.
Cresol (o-)	Higgins, I.J. et al. 1979.	Pentanol (n-)	Higgins, I.J. et al. 1979.
Cyanides	Bohn, H.L. 1977.	Phenol	Schmidt, S.K. et al. 1985.
Cycloheptane	Beam, H.W. et al. 1974.	Phenyldecane (1-)	Higgins, I.J. et al. 1979.
Cycloheptanone	Beam, H.W. et al. 1974.	Phenylnonane (1-)	Higgins, I.J. et al. 1979.
Cyclohexanediol (1,2)	Beam, H.W. et al. 1974.	Phosgene	Turner, N.C. 1973.
Cyclohexanediol (1,3)	Stirling, L.A. et al. 1977.	Propane	Bohn, H.L. et al. 1988.
Cyclohexanediol (1,4)	Stirling, L.A. et al. 1977.	Propene	Dalton, H. et al. 1982.
Cyclohexandione (1,2-)	Stirling, L.A. et al. 1977.	Propylbenzene (n-)	Horvath, R.S. 1972.
Cyclohexane	Stirling, L.A. et al. 1977.	Propylene	Hou, C.T. 1980.
Cyclohexanol	Beam, H.W. et al. 1974.	Putrescine	Bohn, H.L. 1977.
Cyclohexanone	Beam, H.W. et al. 1974.	Pyridine	Dalton, H. et al. 1982.
Cyclohexene	Stirling, L.A. et al. 1977.	Pyrrolidone	Horvath, R.S. 1972.
Cyclohexene oxide	Stirling, L.A. et al. 1977.	Skatole	Bohn, H.L. 1972.
Cyclooctane	Beam, H.W. et al. 1974.	Styrene	Higgins, I.J. et al.1979.
Cyclopentanone	Beam, H.W. et al. 1974.	Sulfur dioxide	Smith, K.A. et al. 1973.
Cymene (p-)	Dalton, H. et al. 1982.	Terpenes	Rasmussen, R.A. 1972.
Decane (n-)	Higgins, I.J. et al. 1979.	Tetrachloromethane	Galli, R. et al. 1989.
Dialkyl sulfides	Fuller W.F. et al. 1983.	Tetradecane	Perry, J.J. 1979.
Dichlorocatechol (3,5-)	Horvath, R.S. 1972.	Toluene	Dalton, H. et al. 1982.
Dichlorodiphenyl		Toluidine (p-)	Higgins, I.J. et al .1979.
methane (p,p'-)	Horvath, R.S. 1972.	Tridecane (n-)	Perry, J.J. 1979.
Diethyl ether	Dalton, H. et al. 1982.	Triethylamine	Fuller W.F. et al. 1983.
Dimethyl disulfide	Oremland, R.S. et al. 1989.	Trichlorobenzoate (2,3,6-)	Horvath, R.S. 1972.
Dimethyl ether	Dalton, H. et al. 1982.	Trichloroethane (1,1,1-)	Galli, R. et al. 1989.
Diphenyl-2,2,2-		Trichloromethane	Galli, R. et al. 1989.
trichloroethane (1,1-)	Horvath, R.S. 1972.	Trichlorophenoxy	
Dodecane (n-)	Perry, J.J. 1979.	- acetic acid (2,4,5-)	Horvath, R.S. 1972.
Dodecylcyclohexane	Beam, H.W. et al. 1974.	Xylene (m-)	Higgins, I.J. et al. 1979.
Ethane	Dalton, H. et al. 1982.	Xylene (o-)	Horvath, R.S. 1972.
Ethanol	Zavarzin, G.A. et al. 1977.	Xylene (p-)	Horvath, R.S. 1972.
Ethylbenzene	Dalton, H. et al. 1982.		

SBR were chosen to bracket those expected to be used in Biosphere 2. Trace contaminants were injected at the beginning of an experiment through the sampling port. Periodically the atmosphere within these systems was sampled and subjected to analysis with a gas chromatograph. To minimize pressure differentials room air was injected into the aquarium to compensate for atmosphere removed.

The most significant finding of our SBR research was the discovery that SBRs are highly variable in their behavior. This is not surprising when one considers the complexity of any natural soil microbial community. We believe, however, that much of the variability of the performance comes not from the soil microbes themselves but rather the environment of the soil and the physical status of the soil bed air purifier. Factors such as rapidly changing soil moisture levels and the methods that soil was placed within a SBR container can result in variable channeling behavior of air through the soil. With channeling, considerable variation in exposure of the soil microbes to the trace contaminants can occur.

Despite the variability we found in SBR behavior the most consistent statistically significant factor in SBR performance was prior exposure to atmospheric contaminants. As shown in a previous figure, the efficiency of removal of contaminants increases with the duration of exposure to a particular contaminant. In the aquaria SBR this was particularly true for ethylene. This graph (Figure 6) shows the increasing efficiency of removal of ethylene over four weeks of exposure. Beginning with a removal rate not different from zero during the first 4 days (the first week is negative due to ethylene production by the soil) the removal of ethylene became essentially total at the end of four weeks. A removal per cent in excess of 100% indicates that the soil bed has removed both the injected ethylene and the ethylene produced by the soil itself.

Figure 3. Soil bed reactors, part of a 72 replicate experimental setup, used in studies on plant growth and development at the Environmental Research Laboratory, University of Arizona.

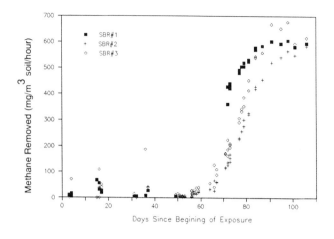

Figure 4. Methane removal in large SBRs as a function of soil type. The graphs also indicate the increased efficiency of removal over time.

The same pattern was noted for propane as displayed in Figure 7. Results for methane, carbon monoxide, and ethane showed similar patterns.

Our hypothesis is simply that exposure to trace contaminants over time allows the growth of microbe populations in the soil that can utilize the contaminants. Anecdotally it appears that these populations can sustain periods of no exposure without significant declines in removal efficiency.

The graph in Figure 8 illustrates that the conditioning effect is observable in soils with inherently less organic matter and lower fertility. In this case unconditioned soil is soil within its first week of exposure to the contaminant gases whereas conditioned soil is the same soil after two weeks of exposure.

Any factor that might promote a larger, healthier population of soil microbes should also improve the scrubbing efficiency of a SBR. Figure 9 shows that when a soil is amended with organic matter (in the form of compost and peat moss) increased scrubbing efficiency should be expected. This graph is a comparison of exposure of the same

Figure 5. Aquaria SBR: 38 liter soil bed reactors used for benchtop tests of air pollutant control.

basic soil to contaminants when amended with organic matter and when left unamended. Clearly the amended soil is more efficient. This implies that soils that support a healthy population of plants would also be more efficient due to the plants' contribution to the soil organic matter within the rhizosphere. Current research should provide a more detailed investigation of this relationship soon.

The last factor I would like to discuss is that of air flow rate through a soil bed air purifier. Ethylene removal was studied as a function of flow rate in one of our early aquaria experiments. The results showed an optimal flow rate of somewhere between two and three atmospheric turnovers/day. This pattern was repeated with the other gases we examined and in our other experiments. While the trend was there this was not a statistically significant result due to the inherent variability of the data (Figure 10). Theoretically however this is not an unexpected pattern due to both enzymatic dynamics and increased channeling at higher flow rates.

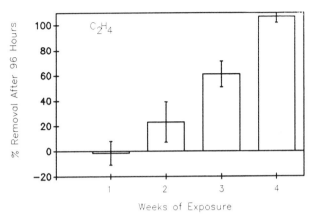

Figure 6. Conditioning effect of exposure to ethylene, a common atmospheric contaminant.

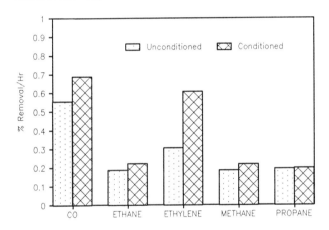

Figure 7. Gas removal in conditioned and unconditioned gray soil.

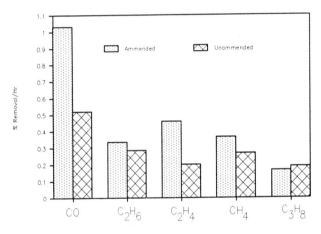

Figure 9. Effect of added organic matter on removal efficiency.

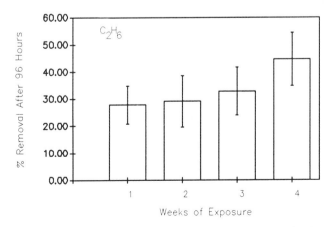

Figure 8. Conditioning effect of exposure to atmospheric contaminants.

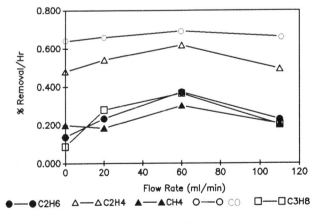

Figure 10. Removal of atmospheric contaminants by a SBR.

When averaged over several experiments the pattern is considerably reduced due to inter-experimental variability.

The last figures deal with the effectiveness of a SBR within a closed system such as Biosphere 2 or any closed system which could be established on another planetary body. During the summer of 1989 we set up a physical scale model of Biosphere 2. This model was to help verify mathematical models of trace contaminant behavior within Biosphere 2. The system consists of two aquaria, one scaled to represent the volume of the Intensive Agriculture Biome (IAB), Habitat, and Lung; and the other scaled to the size of the Wilderness Biomes and its Lung (Figure 11). The total volume of the system is 190 liters. The IAB aquarium has within it a SBR composed of a scaled volume of dirt and a pump to move the atmosphere within this aquarium through the soil. A second pump is lo-cated in the IAB to move air between the IAB aquarium and the Wilderness aquarium. The Wilderness aquarium contains a scaled quantity of soil and vegetation appropriate to the various biomes of Biosphere 2. We also placed a scaled Ocean within the Wilderness Biomes. During our first standardization runs we conducted we found evidence that supported our other research on the utility of SBR. In this experimental work, the removal of representative trace gases was examined when the SBR in the IAB aquarium was operating and when it was not. Figure 12 shows the results of this experiment. Note that for methane (CH_4), ethane (C_2H_6), propane (C_3H_8) and nitrous oxide (N_2O), operation of a SBR substantially reduces their concentrations within the system. Carbon monoxide (CO) seems relatively unaffected by operation of a SBR though this result could be due to the production of CO by the pump when it was operating.

Figure 11. ERL researcher with physical scale model of Biosphere 2 used for soil bed reactor research.

Ethylene concentrations were higher when the SBR was operating than when it was not. This result is probably due to different production rates of ethylene during the two runs. Nevertheless, in both the case of ethylene and carbon monoxide, the atmospheric concentrations of these gases were reduced to less than 20% of their original levels. These data provide the first evidence that a SBR within a closed ecological system would be effective in limiting the levels of atmospheric contaminants.

An analysis of CO_2 production by SBRs revealed that no additional CO_2 is produced when the flow rate of air through a SBR is increased. The regression of the rate of CO_2 production on air flow rate was actually negative, that is, the higher the flow rate of air the lower the rate of CO_2 production. This phenomena is probably due to the effects of increased channeling, and the metabolic depression of the microbial communities due to cooling brought on by evaporation of soil moisture or limitation by soil moisture directly. The initiation of operation of a SBR does however lead to a dramatic increase in CO_2 levels in closed systems. This is due to forcing out the accumulated CO_2 within the soil pores. Continued operation however does not result in higher CO_2 production rates.

ERL, with the support of another group, the Planetary Design Corporation, has also investigated the use of small SBRs for use in office and home environments. This research has indicated that a SBR is also effective in minimizing airborne biological particulates. While the initial operation of a SBR will increase the amount of biological particulates, continued operation of the SBR will reduce the level of fungal spores to quantities less than that noted in a room without a SBR operating.

This research I have presented was conducted for Space Biospheres Ventures to assist in determining the optimal operation of the SBR to be located within Biosphere 2. While it was known in general that SBRs could remove trace atmospheric contaminants, the specific characteristics of SBR performance were unknown. We believe we have made considerable progress in elucidating some of the principles of SBR performance and operation and expect that both our own research and the

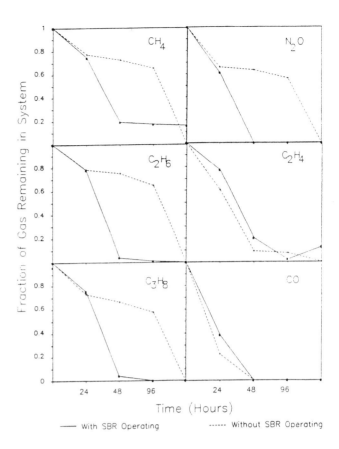

Figure 12. Atmospheric contaminant removal by a soil bed reactor in a closed ecological system.

research conducted by SBV in Biosphere 2 will answer many other questions. SBV has patent applications covering the advances made in SBR technology under this program which have tremendous commercial potential in reducing indoor and outdoor pollution while supporting productive crop or landscape plantings.

ERL is currently working with power generating companies in exploring the methodology of using SBRs and agriculture production for simultaneously reducing CO_2, CH_4, SO_2, and other emissions from power plants and increasing productivity to feed a hungry world. This is just one example of many important interactions between the results of work for Biosphere 1 and 2 benefiting the future success of both.

BioMedical Program at Space Biospheres Ventures

Roy Walford, M.D.
Chief of Medical Operations, Space Biospheres Ventures
Professor of Pathology, UCLA Medical School

There are many similarities and some important differences between potential health problems of Biosphere 2 and those which might be anticipated for a station in space or a major outpost on Mars. We shall not have to deal with microgravity within Biosphere 2, nor with the remote distances of a planetary or even moon base. The demands of time, expense, and equipment would not readily allow medical evacuation from deep space for a serious illness or major trauma, whereas we can easily evacuate personnel from Biosphere 2 if necessary. However, a major albeit self-imposed constraint is to avoid doing so by mistake, i.e., for an illness that could in fact be handled inside Biosphere 2, without breaking closure. Thus, our diagnostic facilities must be first-rate, approaching or fully equivalent to those of a Martian base. Treatment facilities can be somewhat less inclusive, since distance would not compel us to undertake heroic measures or highly complicated surgical procedures on site, and with personnel not fully trained in these procedures.

Now for the similarities between medical requirements of Biosphere 2 and the complex closed ecological systems of biospheres in space or on Mars. The major problems common to all these would seem to be trauma, infection, and toxicity. Handling these requires prompt and effective diagnosis, therapy appropriate to the locale, effective training of personnel, and adequate consultative backup. Regarding this last, we will have computer and high-resolution video communications between Biosphere 2's medical facility and stations at the University of Arizona and UCLA Schools of Medicine. For initial training, a 100-hour "introduction to medicine" course, slanted towards clinical history and physical examination, was given to selected Biosphere 2 personnel by Dr. Dan Levinson of the University of Arizona School of Medicine. This was followed by a week's course in baseline dentistry at the U.S. Naval Hospital in San Diego, Biosphere 2 personnel being permitted to participate in this phase of the Navy's course for Advanced Hospital Corpsmen assigned to isolated stations. Other training is ongoing and will include an intensive course in practical microbiology specifically tailored for Biosphere 2 by the UCLA Hospital Clinical Laboratories.

It is planned that minor and moderate degrees of trauma, including debridement and suturing of wounds, X-ray evaluation of fractures, will be done within Biosphere 2. Portable X-ray equipment and polaroid-like X-ray films (which do not require use of liquid solvents for development) are available. Major trauma will probably be cause for evacuation of the victim(s). Nevertheless, such trauma requires a swift and effective response during the critical first hour, until assistance and evacuation can be mobilized. In short, the Biospherian trauma team must be very good during the first hour. To this end, selected Biosphere 2 personnel are enrolled in the three-day course in immediate (first-hour) management of trauma given by the University of Georgetown School of Medicine, and sponsored by the American College of Surgeons.

We expect bacteriologic and fungal infections, and possibly allergies to pollen or spores, to be the commonest medical problem within Biosphere 2.

The warm, humid, semitropical climate, the rain forest, ocean, savannah, desert and marsh biomes, the agricultural station and animal farm (goats, pigs, chickens), and the daily association of the eight Biospherians with all these areas will assure intimate contact with microbial agents. An atmosphere richer than normal in carbon dioxide will potentiate growth of many of these microorganisms. Of course many human pathogens such as cholera, typhoid, AIDS will not be present at all within our closed space, having been denied entry. However, the rate of evolutionary turnover may well be speeded up within Biosphere 2, with emergence of organisms following mutation/selection, or just the selective pressures of an unusual adaptive stress, which we are not quite accustomed to dealing with in Biosphere 1. For these reasons, microbiology has received considerable emphasis in our program. Using no more than six to eight media and an anaerobic gas pack, we should be able to do primary culture and isolation of all or most organisms that we have to deal with. Primary isolation must be followed by specific identification. This will be done by use of the highly automated Vitek system. The fundamental unit of the Vitek

system is a small plastic plate containing thirty micro-wells, each with a different culture medium. Vitek provides at the moment ten different plates, i.e., 300 different culture conditions, some including antibiotic sensitivities. Besides bacteria, the Vitek system will handle yeast identification, but not fungi. These we propose to identify, at least in part, by more old-fashioned culture and microscopic technics. The same is true for (atypical) acid-fast microorganisms.

So much for trauma and infection. In some ways a stickier problem is the possibility of toxicity in Biosphere 2. There may be offgassing from plastics or other materials, leachates from cement or metal alloys — of no overriding importance outside but dangerous within a totally closed, recycling system. And gases may be locally produced, from composting, for example, or in some instances directly by plants (e.g., ethylene by tomatoes). Many of these agents will be removed by the soil bed reactors inside Biosphere 2. It is not established that all potentially toxic gases can be so removed, and research into this area is part of our present program. Acute toxicity is in one sense the lesser problem because it announces itself with

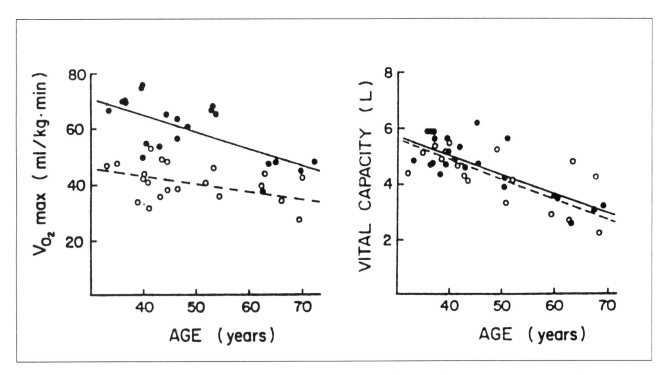

Figure 1. Maximum oxygen consumption and pulmonary vital capacity in relation to age in physically fit (•, —) and sedentary (o, - - - - -) individuals (from Suominen et al., 1980).

obvious symptoms: acute respiratory distress, gastric upset, etc. At least one knows that something is wrong. But poisoning may be insidious, asymptomatic until irreversible damage has been done, for example to bone marrow, liver, or possibly brain, and once started may continue to be progressive even if the patient is removed to a non-toxic environment.

Frequent quantitative analyses of blood indices, and qualitative (microscopic) analyses of blood morphology, with bone marrow aspiration if indicated, may help detect early signs of hematologic injury. Developing liver injury may be foreshadowed by altered blood chemistry, particularly selected enzymes. To estimate these changes but avoid the self-defeating use of organic solvents in the methodology, we shall employ Eastman Kodak's Ektachem system for dry reagent chemistry. Like Vitek, this is a compact system. At the moment 28 different blood chemistries, including enzymes, protein, glucose, bilirubin, the electrolytes, cholesterol, and lipoproteins can be measured accurately. Complete reagents for each test are contained in dry state on a small square about the same area and thickness of a quarter. A large number can thus be stored in Biosphere 2 before closure.

Because of the physically closed, electronically open nature of Biosphere 2, inside personnel must be capable not only of using but of repairing the above various equipment items. Training in these aspects is ongoing with the various parent companies.

I want to branch off now into ways of monitoring health, besides doing these various above-mentioned tests. One of the keys to that actually comes from gerontology. Gerontologists have been concerned with monitoring age specific biomarkers in humans and have developed a substantial battery of tests to that end, with the goal of measuring "functional age" as opposed merely to chronological age. These include, for example, vital capacity, maximum work rate, suppressor cell response, presence or absence of autoantibodies, delayed type hypersensitivity, serum albumin and globulin levels, reaction time, tapping time, hearing threshold at a fixed frequency, plus others (Weindruch and Walford, 1988). A few of these are illustrated

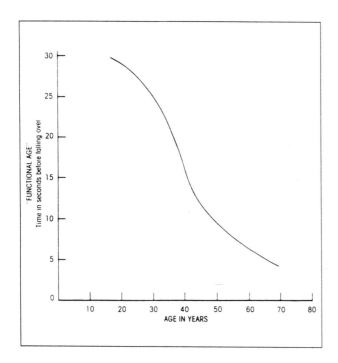

Figure 2. Static balance as a biomarker of age. Close eyes, stand on one leg (left if you are right-handed), don't move foot. How long before you fall over? Score = average of 3 trials. (From Walford, 1986.)

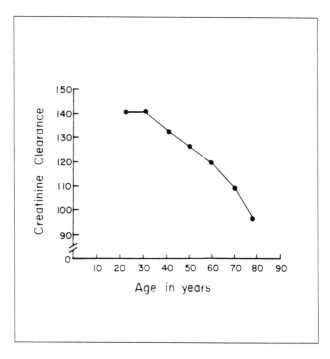

Figure 3. The effect of age on the important measure of kidney function known as Creatinine Clearance (adapted from J.W. Rowe et al., Journal of Gerontology, 31:155, 1976).

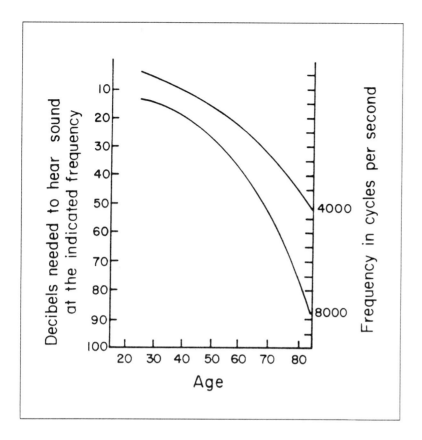

Figure 4. Auditory threshold in decibels in relation to age in women, at sound frequencies of 4000 and 8000 cycles per second (adapted from J.F. Corse, in *Lectures on Gerontology, Vol. 1, part B., Biology of Aging,* ed. A. Viidik [New York: Academic Press, 1982], p. 441).

1. Vibrotactile sensitivity
2. Memory
3. Vital Capacity
4. Forced expiratory volume-1
5. Alternate button tapping time
6. Highest audible pitch
7. Visual accommodation
8. Auditory reaction time
9. Visual reaction time(VRT)
10. Muscle movement speed (MMS)
11. VRT with decision
12. MMS with decision

Figure 5. Physiological functions measured automatically by H-Scan (Hoch Company, 2915 Pebble Drive, Corma del Mar, CA 92625).

in Figures 1, 2, 3, and 4. Automated equipment for measuring some of these is available on a commercial basis (see Figure 5).

The biomarker approach is quite applicable to health assessment in Biospherians and, I suggest, also in astronauts and cosmonauts. I understand from personal conversations with some of the astronauts, including some who have reached retirement age, that most of them return every year to run through a large battery of physiologic tests, but as far as I know very little is being done with this data. The data should be quite susceptible to biomarker analysis according to technics worked out by gerontologists.

BIBLIOGRAPHY

1. Weindruch, R., and Walford, R. L.: The Retardation of Aging and Disease by Dietary Restriction. Charles C. Thomas, Springfield, 1988.

2. Suominen, H., Heikkinen, E., Parkatti, T., Forsberg, S. and Kiiskinen, A.: Effects of lifelong physical training on functional aging in men. Scand. J. Soc. Med. 14:225, 1980.

3. Walford, R. L.: The 120-Year Diet. Simon and Schuster, New York, 1986.

4. Rowe, J.W., Andres, R., Tobin, J.D., Norris, A.H., and Shock, N.W.: The effect of age on creatinine clearance in men: a cross-sectional and longitudinal study. J. Gerontol. 31:155, 1976.

The NASA CELSS Program

Maurice M. Averner, Ph.D.
Program Manager, NASA CELSS and Biospherics Programs,
Life Sciences Division, NASA Headquarters, Washington D.C.

OVERVIEW

The NASA Controlled Ecological Life Support System (CELSS) program was initiated in 1978 by the Life Sciences Division, Office of Space Science and Applications (OSSA), with the premise that NASA's goals would eventually include extended-duration missions with sizable crews requiring capabilities beyond the ability of conventional life support technology. Currently, as mission duration and crew size increase, the mass and volume required for consumable life support supplies also increase linearly. Under these circumstances the logistics arrangements and associated costs for life support resupply will adversely affect the ability of NASA to conduct long-duration missions. A solution to the problem is to develop technology for the recycling of life support supplies from wastes. The CELSS concept is based upon the integration of biological and physico-chemical processes to construct a system which will produce food, potable water and a breathable atmosphere from metabolic and other wastes, in a stable and reliable manner. A central feature of a CELSS is the use of green plant photosynthesis to produce food, with the resulting production of oxygen and potable water, and the removal of carbon dioxide.

The development of an operational CELSS will provide economic, psychological and mission operations benefits. For long-duration missions, such as permanent lunar or Mars bases, where logistics supply is very costly or impractical, the development of a full integrated bioregenerative life support system will be enabling. As the duration of future manned space missions increases, a cross-over point is reached where it will be more economical to provide life support supplies by the recycling of metabolic and hygiene wastes than to incur the repeating costs of resupply. In-situ regeneration of life support consumables will protect the mission from unpredictable and potentially disastrous interruptions in the logistics train.

The development of bioregenerative life support systems should be viewed as a key enabling step in NASA's ability to support humans for long durations in space. Such a system will have economic benefits, radically lowering costs of mission life support, mission operations benefits by substantially reducing the need for consumables that must be resupplied or brought along, and psychological and health benefits, by providing astronauts with a dependable supply of fresh food from a self-contained system.

TECHNICAL APPROACH

The general approach to CELSS research and development activities is to accomplish successive stages of prototype system development, based upon and supported by appropriate ground-based and flight experiments, so that the development of operational space systems can begin soon after the turn of the century. A CELSS can be viewed as an integrated set of biological and physico-chemical subsystems, functioning through processes of regeneration of recycling to sustain human life.

These major subsystems include:

1. Biomass production (plant and secondary animal production)

2. Biomass processing (food production from biomass)

3. Water purification

4. Air revitalization

5. Solid waste processing

6. System monitoring and control

These subsystems are interactive and interdependent. Research needs include both ground-based and flight studies that range from determining the environmental requirements for optimal plant productivity and the effects of micro-gravity on plant growth, to the problems inherent in the development of the technology required for the recycling of human and plant wastes. The development of these subsystems, their integration, and the characterization of mission-specific CELSS variants will be carried out by a series of projects as described below.

PROGRAM DESCRIPTION

The CELSS program is structured around six major elements, each of which represents a major area of science and technology research and development. These elements are:

1. Research Program

A continuing program to develop advanced component technologies for CELSS projects and provide scientific support for the development of biologically-based processors. Activities include the development of physical-chemical waste processing techniques, food processing scale-down, and development of advanced lighting systems.

2. Systems Integration and Control

Directed at the design, development, testing and evaluation of models and laboratory- scale experimental systems bearing on CELSS system monitoring, control and behavior. Activities include the development of system and process models, and an interactive program of systems testing under laboratory conditions.

3. Breadboard Project

Ground-based project at the Kennedy Space Center which will determine if lab-scale plant growth, food production and waste processing techniques can be successful when tested at an operational scale. Does not include humans in the system. The completion of the Breadboard Project will be a major step in the demonstration of CELSS feasibility.

4. Human-rated Test Facility

Ground-based project which will provide a full-scale test of a complete CELSS, including all biological and physical chemical systems and crew interfaces. Based upon current and anticipated experience with the Breadboard Project and planned to be operational in the middle 1990's.

5. Advanced Mission Concept Studies

Directed at developing mission specific options for CELSS applications for the suite of potential future manned missions such as lunar and Mars bases.

6. Space Flight Experiments

A program for determining the productivity, adaptability and stability of food crop plants and their supporting systems in a microgravity or reduced-gravity environment.

THE CELSS BREADBOARD PROJECT: PLANT PRODUCTION

William M. Knott, Ph.D
Director, NASA CELSS Breadboard Project
Kennedy Space Center

INTRODUCTION

I will describe NASA's Breadboard Project for the CELSS program. For those familiar with the Breadboard Project at Kennedy Space Center, it should bring you up to date on what has happened during the last year; for the others, it will be a short introduction to the project.

The simplified schematic of a CELSS is shown in Figure 1. I start with the schematic to emphasize that we are taking a modular approach to constructing the CELSS Breadboard. We are researching each module in order to develop a data set for each one prior to its integration into the complete system. I will concentrate on the data being obtained from the Biomass Production Module or the Biomass Production Chamber. The other two primary modules, food processing and resource recovery or waste management, will be discussed only briefly. The crew habitat module will not be discussed at all during this presentation.

The primary goal of the Breadboard Project is to scale-up research data to an integrated system capable of supporting one person in order to establish feasibility for the development and operation of a CELSS. Breadboard is NASA's first attempt at developing a large scale CELSS. Research emphasis in our work over the past three years has been on the Biomass Production module. In late 1990 integration of the food processing and resource recovery modules will be initiated. The goal is to have a complete functional system operational by 1993. The crew habitat module will only be simulated during the Breadboard Project.

BIOMASS PRODUCTION MODULE

Biomass Production Chamber:

The Biomass Production Chamber (BPC) is a 7.5 meter tall by 3 meter in diameter stainless steel cylinder (Figure 2). This cylinder or chamber is oriented in the vertical position and has an internal volume of 113 m^3. The chamber itself was used for leak testing of capsules during the Mercury spaceflight program. We renovated and modified it so that it could be used as a large atmospherically sealed plant growth chamber.

The chamber is divided in half by a floor making it a two story structure. An extensive air distribution and conditioning system was added to the outside of the chamber. Eight racks were built and installed on each of the two floors in the chamber. Each stainless steel rack has two light banks with a shelf under each to accommodate plants during their growth. Air flow in the chamber is across the plant canopy and back through the light banks into the duct system. Lighting in the chamber is by high pressure sodium lamps and at full intensity is approximately one half full sunlight. Environmental control for each floor or compartment of the chamber is separate. A steel platform was built around the chamber in order to allow access to the chamber and to the ducting around the outside.

In the control room for the BPC is housed a microprocessor that is programmed through a computer station to control conditions in the chamber. A fundamental principle followed in construction of the control and monitoring system was that the control system would be separate from the monitoring system. Therefore, each system has its own sensors and computer. The primary components controlled and monitored from this room are nutrient delivery, environmental parameters and atmospheric gases. All data collected are stored in a central main frame computer.

All data collected can be displayed on any computer in the facility in both graphic and tabular form. Digital displays in the control room give current readings for any parameter being measured in the chamber. Visual and auditory alarms are activated when any parameter goes out of range during chamber operation. The interior of the chamber is under constant surveillance by television cameras. One camera on each floor has a pan-tilt-zoom capability which allows one to inspect for leaks or

other problems in the chamber and to make close-up observations of the plants from outside the BPC.

The atmospheric gas system can control and monitor up to four gases. Currently we are monitoring oxygen and carbon dioxide and are controlling carbon dioxide. Gas control is accomplished by the introduction of the appropriate gas from pressurized cylinders located outside the chamber. A system of valves and switches in the gas racks allows control of gases at the requisite levels. Monitoring of trace gases is accomplished through gas chromatography and mass spectrometry of samples taken from the chamber. Additional gas control and/or monitoring capability will be added to the BPC as requirements are identified.

The nutrient delivery system is another major component controlled and monitored in the BPC. This system is made up of four large nutrient solu-

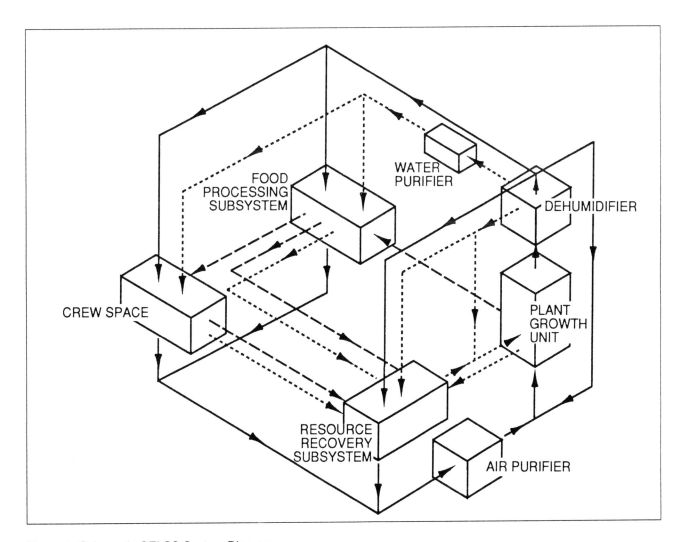

Figure 1. Schematic CELSS System Diagram

tion holding tanks outside of the chamber and 64 plastic plant growing trays inside of the chamber. The 64 growing trays are divided between four levels, 16 trays per level, with each level receiving solution from one of the four storage tanks. All plants are grown in thin film hydroponics. Nutrient solution is delivered to the back of each tray, flows across the bottom of the tray, and returns to each nutrient tank through a common guttering system. This system is obviously dependent on gravity for its operation. We monitor flow rate, pH, conductivity, and liquid level for each of the four tank systems. Samples are removed periodically from each tank so that inorganic chemical and microbial analyses can be conducted. Currently, pH and liquid level are the only parameters being actively controlled in this system.

Wheat Productivity Test:

We have conducted several trials of wheat in the Biomass Production Chamber. The crop growing area for each level in the chamber is approximately 5 sq meters which makes the total growing area of the chamber approximately 20 sq meters. Wheat is the first crop on which we have completed tests in the BPC. These tests have concentrated, as will future research, on measuring mass flow through, energy input to, and contaminant buildup in the system. During each test, we are continually monitoring CO_2, oxygen and water in the system in order to determine flow rates through the plant canopy. Energy input is also measured so that the demands of the system can be determined. All crop tests in the BPC are conducted from the seed stage to full plant production.

Carbon dioxide is continuously monitored through each test. It is controlled at 1000 ppm during the test period. The 1600 ppm peaks that show up periodically occur when the lights are off in the chamber. Such fluctuations in carbon dioxide due to the presence or absence of active photosyn-

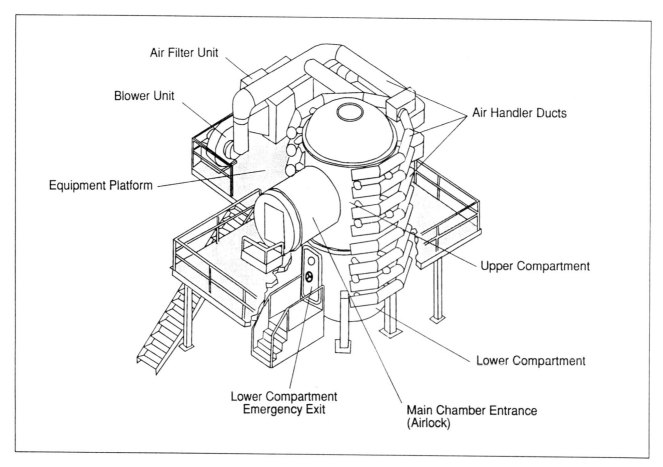

Figure 2. The Biomass Production Chamber.

thetic activity must be taken into consideration when one is designing a CELSS. We have measured the rate of carbon dioxide uptake in the chamber when there is a full canopy of photosynthetically active plants. During these trials, the chamber's carbon dioxide level is elevated to a set point, then the valve controlling CO_2 input into the chamber is closed. The rate that the carbon dioxide is drawn out of the chamber indicates the photosynthetic activity of the crop under the existing environmental conditions. One can change parameters such as temperature and irradiance levels in the chamber during these tests and observe the corresponding changes in photosynthetic rate. We have data on the amount of CO_2 used on a daily basis throughout an entire wheat life cycle. We have examined photosynthesis and respiration data for a mature crop of wheat on a meter sq per second basis. Manipulating temperature and irradiance levels impacts photosynthesis and/or respi-

ration rate in the mature wheat canopy. One could utilize these effects, for example, in regulating a CELSS for optimum uptake of carbon dioxide. We have also determined the light compensation point for this canopy of wheat, the total uptake of carbon dioxide by the wheat in the chamber on an hourly basis and how carbon dioxide levels influence transpiration rates. All gases added to the BPC during tests are metered in through mass flow valves.

During all crop tests there is a set of environmental parameters that are constantly measured and recorded. During the wheat trials these included irradiance levels, relative humidity, temperature and atmospheric pressure which are routinely measured during each test of a crop. Parameters measured in the nutrient delivery system include flow rate, liquid level, pH and conductivity. In addition, the amount of condensate water collected from each of the two compartments on a daily basis is measured and recorded during each

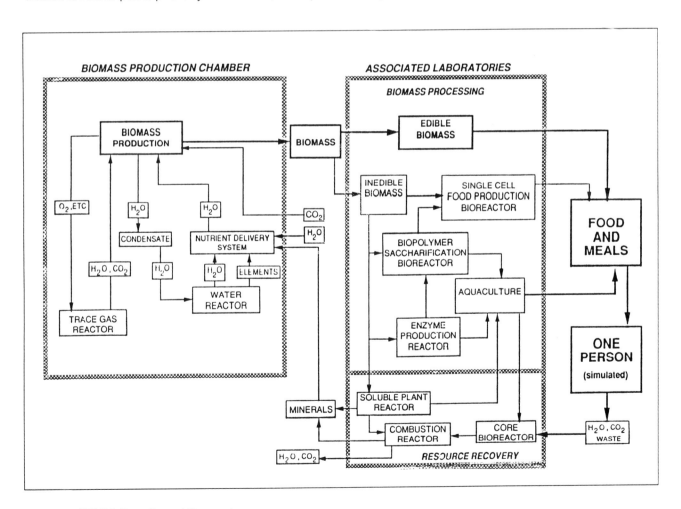

Figure 3. CELSS Breadboard Concept.

trial. The composite of these environmental data, the photosynthetic gaseous exchange data presented previously, and the measurement of biomass production allows one to begin to understand the operational requirements of a Biomass Production Module, at least for wheat.

When measuring mass flows through a system and determining rates of contaminant buildup, one must continually measure the leak rate of the facility being used. A decay curve for carbon dioxide over a 48 hour period of time from an empty but operating Biomass Production Chamber shows a slow decline of carbon dioxide which when mathematically analyzed allows a determination of the chamber leak rate. The best leak rate which we have measured is 2.5% of the chamber volume leaked per day. The average leak rate for the operation of the Biomass Production Chamber is approximately 5% of the volume leaked per day. We are continually sealing the chamber during each operation to improve our atmospheric leak rate.

CANDIDATE CROP SPECIES

A variety of crop species are currently being prepared for testing in the BPC. These tests are being conducted in conventional plant growth chambers located within the Life Sciences Support Facility at Kennedy Space Center. The next crop to be tested in the Biomass Production Chamber will be soybean. Preliminary research on this crop has centered on response of the soybean to various irradiance levels and elevated carbon dioxide concentrations. Potatoes will be studied in the BPC next year. We have already grown two varieties in thin film hydroponics. Both white potatoes and sweet potatoes have formed tubers and storage roots, respectively, in the hydroponic system. Other plant species being prepared for testing in the BPC include: peanuts, lettuce, radishes, tomatoes, sugar beets, bush beans, and rice. Data generated on each of these crop species by the research program will be utilized in preparation for growing these plants in the Biomass Production

TASK	ACCOMPLISHMENT				
	FY 89	FY 90	FY 91	FY 92	FY 93
CROP PRODUCTION	Wheat	Wheat, Soybean	Wheat, Soybean, Potato	Multiple Crops	Continuous Production
WATER RECYCLING	Design	Fabricate And Install	Operate	Modify	Operate
ATMOSPHERIC GAS CONTROL	Measure And Analyze	Design and Fabricate	Install	Operate	Operate
BIOMASS PRODUCTION CHAMBER					
ADJACENT LABORATORIES					
BIOMASS PROCESSING	Harvest, Store Measure, Analyze	Establish Laboratory	Process Edible	Process Inedible	Evaluate
RESOURCE RECOVERY	Store, Measure Analyze	Establish Laboratory	Recycle minerals, Design Combustion System	Install	Evaluate
DATA MANAGEMENT	Environmental Nutrient Delivery, Atmospheric	Energy, Trace gas	Multivarient Analysis	Models	Validate

Figure 4. CELSS Breadboard Project Matrix.

Biological Life Support Systems

Chamber. Current plans are to test at least five crop species in the BPC by the end of 1993.

SYSTEM INTEGRATION

Figure 3 shows a block diagram of the initial CELSS that we plan to develop and operate during the Breadboard Project. Final integration and initial testing of this system is scheduled for 1993 and 1994. The left hand side of this illustration is of the Biomass Production Chamber. Data being collected from the operation of this chamber include: biomass, amounts of condensate water, elemental uptake, carbon dioxide and oxygen fluxes, microbial constituents, concentrations of trace organics in the atmosphere, and presence of trace organics in the nutrient delivery solution. We are also collecting information on manpower requirements, energy use, spare parts requirements, and operational reliability. The condensate water loop on the BPC will be closed during 1990 and design completed on a trace gas control system if one is required.

The right side of Figure 3 illustrates the components of the resource recovery, biomass conversion, and food processing modules to be incorporated from 1990 through 1992. These functions will be conducted in laboratories adjacent to the BPC. Analytical chemistry and microbial diagnostic laboratories will also be included in this space. Food processing activities will concentrate on producing a variety of meals from a few crop species. Equipment required to process the edible plant material will be developed and/or tested in conjunction with BPC operations. Biomass conversion activities will concentrate on cellulose conversion of the inedible part of the plant biomass. Subsystems to be tested for this conversion include: enzymatic digesters, single cell protein reactors, and aquaculture. Resource recovery activities will concentrate on the conversion of the final unused material in the system into an acceptable nutrient solution for the plants. Subcomponents to be tested in this effort include: a leachate reactor for the plant biomass, a microbial reactor, and an oxidation/combustion reactor as a final element. We are currently conducting some

research into the development of these modules. All resource recovery, biomass conversion, and food processing components will be functionally integrated with the BPC operations. Each subcomponent will be sealed as required to develop a mass flow database. Data required to determine system operations for each component including mass flow, energy use, and operational reliability will be collected during all trials. Trials of at least a six months duration will be conducted when the total system is functional. The expected activities to be completed during the next four years are summarized in Figure 4.

SUMMARY

The plant production module (Biomass Production Chamber) of a initial CELSS is currently in operation. Data required to establish the mass flow of carbon, hydrogen, oxygen and nitrogen through this system along with information on energy use and operational requirements are currently being collected. The construction of laboratories to accommodate the resource recovery, biomass conversion, and food processing modules of a CELSS is nearing completion. At least 5 crop species will be tested in the BPC by the end of 1991. All subcomponents of the resource recovery, biomass conversion, and food processing modules should be developed and tested by the end of 1993. Initial feasibility testing of a complete CELSS should be completed during the 1993-1994 time frame. The integration and testing of this complete system will generate numerous questions and problems that will require research to solve. This initial testing of a CELSS is the first step in an iterative process that will ultimately produce a functioning CELSS. Many areas will require the development of basic scientific data and/or new technologies prior to the use of a CELSS for life support during long duration space flight. Research and development of a CELSS will require many years of very intensive research and development. Therefore, these initial efforts must be started now if we ever hope to reach our ultimate goal, the permanent presence of humans in space.

CELSS Research and Development Program

David Bubenheim, Ph.D.
Advanced Life Support Division
Regenerative Life Support Branch
NASA Ames Research Center

A Controlled Ecological Life Support System (CELSS) will be a regenerative system which incorporates biological, physical and chemical processes to support humans in extra-terrestrial environments. The key processes in such a system are photosynthesis, whereby green plants utilize light energy to produce food and oxygen while removing carbon dioxide from the atmosphere, and transpiration, the evaporation of water from stomata. Development of a CELSS requires identification of the critical requirements that will allow the system to operate with stability and efficiency. Identifying and meeting those requirements will be accomplished through scientific experimentation and technology development on the ground followed by space flight testing to validate microgravity and reduced gravity adaptability of the system.

NASA's Ames Research Center (ARC) has responsibility for three major CELSS program elements:

1) Research and Development (R & D)

2) System Integration and Control

3) Space Flight Experiments

The Research and Development Program includes evaluation of new ideas and development of advanced principles and technologies in the areas of biomass production, waste processing, water purification, air revitalization and food processing. System Integration and Control involves identification of how the individual component processors of a CELSS can be linked and managed to operate in concert as a system. Both the R&D and System Integration and Control program elements rely on the long- term involvement and interaction of NASA and university scientists and engineers. The Space Flight Program currently is planning for the CELSS test and demonstration hardware to be included as part of Space Station Freedom.

RESEARCH AND DEVELOPMENT

Approach. The CELSS program goal is development of a life support system based upon combining biological and physical/chemical processes capable of recycling the food, air and water needed to support long-term missions with humans in space. Efficiency of the system will be determined based on the ability of the system to recycle mass, thus reducing or eliminating resupply, and the production of human usable products (food, water, O_2, and CO_2 removal) per unit input to the system. The inputs considered important to CELSS system efficiency are volume, energy, time and mass. While these inputs are clearly important, the relative importance of each is subject to change based on mission scenario. A CELSS or individual component technologies may have application in a range of mission scenarios including lunar and planetary bases, space stations and planetary transit.

CELSS research and development has concentrated on characterizing operation of the potential component technologies. For the plant system, the approach has been to identify the flexibility and response time for the food, water and oxygen production, and carbon dioxide consumption processes. To deal with the possibility of changing input limitations, depending on mission scenario, response surfaces are being developed to characterize system performance as a function of inputs. These response surfaces will be utilized to develop potential system designs for specific mission scenarios. Input limitations can be identified for each mission and the response surfaces will feed system trade studies to determine product priority and optimum system design, a process referred to as constrained optimization.

Plant Research. The goal of the R&D program in plant/crop physiology is to characterize the ability of a plant-based system to provide food, O_2, purified water and remove CO_2 from the closed environments of spacecraft for the purpose of life support. The emphasis of plant research to date has been placed on food production with particular attention to methods of reducing the crop area (volume) required to sustain a human, compared with the area presently required in terrestrial agriculture. The discipline of crop physiology has been invoked with the aim of understanding the dynamics of yield development. Crop physiology is related to ecology but without the competition of diverse species and follows a biographical approach to crop development, with emphasis on the critical stages in yield determination and controlling factors for each stage. Research and development includes the conduct of basic research at universi-

Figure 1. Controlled environment plant growth chambers provide control of radiation quality and quantity, carbon dioxide level, humidity, hydroponic nutrient solution. (Photo: NASA Ames Research Center)

ties and at Ames, using controlled environment plant growth chambers, including control of radiation quality and quantity, CO_2 levels, humidity, and hydroponic nutrient solution delivery (Figure 1), and the conduct of closed systems research and study utilizing the Crop Growth Research Chamber (CGRC) at Ames (Figure 2).

The basic research program includes cooperative agreements with university investigators and performance of research at Ames. The goal of the basic research program is to characterize the performance of crop plants and identify optimum environments allowing full expression of the genetic potential (including nontraditional systems of algae, bacteria, and yeast). Studies related to plant purification of concentrated liquid waste streams, as well as, polishing of more dilute waste streams such as hygiene and grey water are also pursued. Nutrients derived from waste streams (recycled) via waste processing will be evaluated and acceptability for plant growth determined.

Crops included in the research were selected for specific purposes. Wheat was selected as a carbohydrate source with the canopy architecture of a grass, potato as an alternative carbohydrate source with a broadleaf canopy architecture, soybean because of the relatively equal proportions of carbohydrate, proteins, and fats, and lettuce was selected as a model photosynthetic system which is not complicated by monocarpic senescence. For each of the crops selected there already exists a large body of knowledge concerning genetics, productivity and response to the environment. Building on present knowledge, environmental manipulation has been practiced in attempts to achieve maximal production in these model crop systems. Light quality, quantity, and periodicity, temperature, nutrient solution delivery and quality, CO_2 concentration in the atmosphere, plant density, and other factors have been altered from traditional agricultural systems to increase productivity.

Accomplishments over the past several years include: exceeding world record field yields, reducing seeding to harvest cycles by more than 50%, improving light utilization efficiency by a factor of 4, proving feasibility of a crop based CELSS where

Parameter	Control Range	Accuracy
Atmospheric Environment		
Air Temperature	5-40°C	±1°C
Air Pressure	+ 15 mm H_2O(gage)	± 5 mm
H_2O(gage)		
Relative Humidity	35-90%	±2%
Air Composition		
Nitrogen	75-95%	±5%
Oxygen	5-25%	±5%
Carbon Dioxide	25-5000 $\mu mol\ mol^{-1}$	±0.2%
Air Flow Rate	0.1-1.0 $m\ s^{-1}$	±0.1 $m\ s^{-1}$
Photosynthetic Photon Flux	0-3000 $\mu mol\ m^{-2}\ s^{-1}$	±2%
Hydroponic Environment		
Temperature	5-40°C	±1°C
pH	4.8-8.0	±0.2 units
Conductivity	0.5-5.0 $dS\ m^{-1}$	±0.5 $dS\ m^{-1}$
Oxygen Concentration	5-20 $\mu mol\ mol^{-1}$	±2%

Table 1. CGRC Science Requirements for Environmental Control.

10 to 15 m^2 of crop can provide the food energy required to sustain one person and produce oxygen and water in great excess of their needs.

Current efforts are in areas of improving efficiency of the cropping system even more, evaluation of potential for phasic manipulation of plant development to further reduce the time to harvest, evaluation and selection of "new" CELSS crop plants appropriate for a balanced human diet, manipulation of plant metabolism to reduce production of inedible biomass, and increase harvest index (edible biomass / total biomass).

Waste Processing. The major objective of the waste processing R&D program is to evaluate, develop and select candidate physical, chemical and biological waste treatment technologies for processing and recycling wastes. The evaluation and selection process includes pre- and post-treatment technologies that are needed for waste processing. Subsystem evaluations include material and energy balances and development and validation of models. Research in technology development is conducted at Ames and in cooperation with university investigators as appropriate.

Past work has specifically emphasized characterization of waste stream quality and quantity in present Space Shuttle missions and proposed missions including CELSS. Potential methods for processing individual waste streams to usable forms are identified by coupling stream constituents with

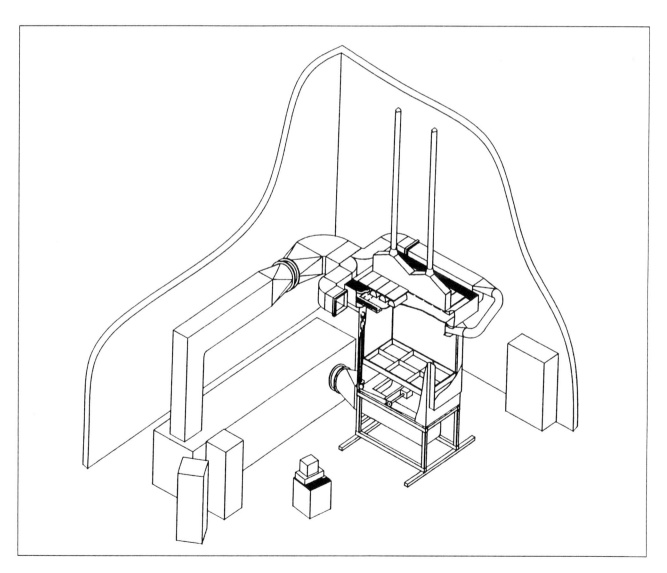

Figure 2. Anticipated physical appearance of the Grop Growth Research Chamber with chamber cut away to chamber interior with view of root zone compartments of the hydroponic system. (NASA Ames Research Center)

a desirable process product. Significant effort has been spent on developing oxidation processes, in particular wet oxidation and super-critical water oxidation. The products of these oxidation treatments, water, CO_2 and inorganic salts, are all desirable in a plant based life support system. Commercially available technologies such as incineration are also being evaluated.

Closed Systems Studies. The Crop Growth Research Chamber (CGRC) is used for the study of plant growth and development under stringently controlled environments isolated from the external environment (closed) and is designed for the growth of a community of crop plants (Table 1). The CGRC is the individual unit where various combinations of environmental factors can be selected and the influence on biomass, food and water production and O_2/CO_2 exchange of crop plants are investigated (Figure 3). Several Crop Growth Research Chambers and laboratory support equipment provide the core of a closed systems plant

research facility. This facility will be utilized for research, technical studies (development and evaluation of technology), system control, system modeling (development and validation), and system operation. Biomass produced in the CGRC and other controlled environment facilities at Ames will be made available for testing in the waste processing systems.

The closed systems plant research facility will supply a defined operation scenario for the plant component of the integrated experimental regenerative system and operate concurrent with integrated system evaluation.

System Control and Integration

Operation and control of a stable system is essential for development of a reliable life support system. The crop growth unit is only one portion of a CELSS but the crop plants function as several unique component processors. Carbon dioxide is removed from the atmosphere while oxygen is

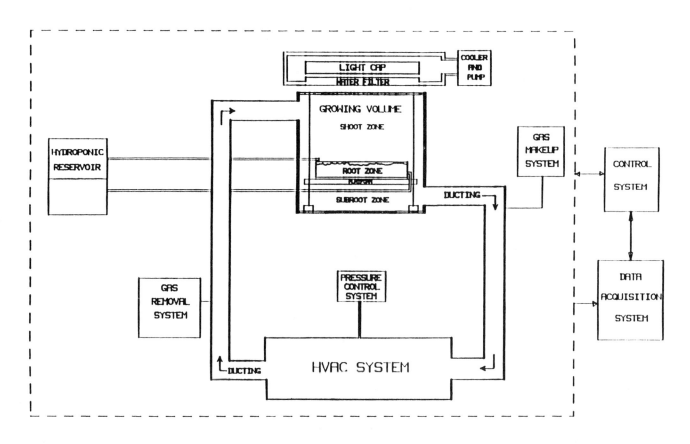

Figure 3. Block diagram of component subsystems and physical zones of the Grop Growth Research Chamber. (NASA Ames Research Center)

introduced through photosynthesis. Plant transpired water has been filtered through uptake by the root and incorporation of solutes into tissue before being evaporated from the interior of stomata of the leaves to the atmosphere. Transpiration rate can be manipulated over a wide range by environmental conditions. Carbon dioxide utilization and oxygen and water production are dynamic systems with short response times and the rate at which these processes operate can be varied as needs for a particular product vary. Of course food is being produced by the plants at the same time; the response time for expression of perturbations in the food production process is much greater than that observed for the other plants processes.

Edible plant yield is the integration of development during several unique phases between germination and harvest. Understanding the dynamics of yield development, i.e. having knowledge of crop responses to environmental manipulation during yield critical phases, is essential to predicting system performance. Carbon dioxide uptake, and oxygen, water, and food can all be considered as products of the plant component of a CELSS. Information required for trade-off analysis to determine the short- and long-term gains and losses resulting from environmental manipulation during the life cycle of a crop as required for the desired plant product will be provided.

Future interface with candidate unit processors on a laboratory scale will be possible. As candidate processes are developed for such operations as waste processing, oxygen removal and storage, nutrient recycle, and harvest and food processing, laboratory scale prototype units could be interfaced with the CGRC. Performance of these processors and requirements for interface with a crop growing unit could be evaluated.

Integrated regenerative systems evaluation involves selection, integration and operation of technologies and subsystems developed in plant production and waste processing R&D programs

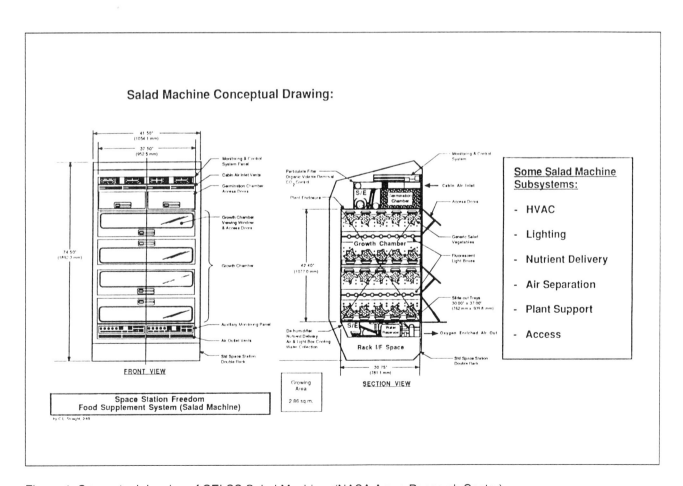

Figure 4. Conceptual drawing of CELSS Salad Machine. (NASA Ames Research Center)

in concert as part of an operational regenerative system. A CGRC (or modified CGRC) will be utilized as the plant growth component of the experimental regenerative system. Waste management subsystems will be sequentially added to the CGRC. Interface requirements for physical, chemical and biological subsystems will be defined. Technical development for automation of functions such as planting, harvest, food processing and conditioning treatment of waste streams before or after processing will be accommodated. CELSS system models and control strategies will be tested for the first time in real closed loop systems, mass and energy balances will be determined and the dynamics of the CELSS system defined for various input limitations (including those imposed by mission scenarios). The ultimate goal of integrated systems evaluation will be design specifications for the crew scale (possibly human rated) life support system testing.

Flight Test and Experimentation

The major emphasis for space flight has been planning for the CELSS Test Facility and the Salad Machine, both to be operated on Space Station Freedom. The CELSS Test Facility (CTF) is part of the NASA Life Sciences Space Biology Initiative. Capability for production of several generations of plant communities and the study of microgravity effects on plant performance is the goal of the CTF. The Salad Machine is being designed to regularly supply crew members of Space Station with salads (Figure 4). Precursor missions on shuttle to test nutrient delivery, germination and transpired water recovery systems for CTF and Salad Machine are being planned.

SUMMARY

Research in Controlled Ecological Life Support Systems conducted by NASA indicate that plant based systems are feasible candidates for supporting humans in space. Ames Research Center has responsibility for Research and Development, System Integration and Control, and Space Flight Experiment portions of the CELSS program.

Important areas for development of new methods and technologies are biomass production, waste processing, water purification, air revitalization and food processing. For the plant system, the approach has been to identify the flexibility and response time for the food, water and oxygen production, and carbon dioxide consumption processes. Tremendous increases in productivity, compared with terrestrial agriculture, have been realized. Waste processing research emphasizes recycle (transformation) of human wastes, trash and inedible biomass to forms usable as inputs to the plant production system. Efforts to improve efficiency of the plant system, select "new" CELSS crops for a balanced diet, and initiate closed system research with the Crop Growth Research Chambers continue. The System Control and Integration program goal is to insure orchestrated system operation of the biological, physical, and chemical component processors of the CELSS. Space flight studies are planned to verify adequate operation of the system in reduced gravity or microgravity environments. The CELSS program's objective is to provide the technology required to support human life during NASA's future long duration missions.

Plants and their Microbial Assistants:
Nature's Answer to Earth's Environmental Pollution Problems

B.C. Wolverton, Ph.D.
Director, Wolverton Environmental Services
Picayune, Mississippi

Before my recent retirement from the U.S. Government, I was employed with NASA as a research scientist at the Stennis Space Center in Mississippi for over 18 years. These past 18 years have been a challenge to maintain funding for continuing research in the utilization of higher plants and their associated microorganisms to solve environmental pollution problems on Earth and in future space applications.[1-12] Hopefully, this research with plants and microorganisms will continue at NASA. Even though I have retired from NASA, I will continue my research and will concentrate on applying this technology to solving some of Earth's environmental pollution problems.

If man is sealed inside closed facilities, we all know he becomes a polluter of the environment. It is also common knowledge to most people that man cannot survive on Earth without green photosynthesizing plants and microorganisms. Therefore, it is vitally important that we have a better understanding of the interactions of man with plants and microorganisms. (Figure 1) Biosphere 2 and other related studies presently being conducted or planned, hopefully, will supply data that will help save planet Earth from impending environmental disaster.

I personally feel that a promising solution to the Earth's environmental pollution problems is the development of a means to utilize both air and water pollution as a nutrient source for growing green plants. To this goal, I have dedicated the past twenty years of my life. As I tour the world and lecture on this approach to environmental pollution control, people are beginning to understand and accept the idea of using nature to clean our environment.

Sewage is now being used as a nutrient solution for growing plants while the plant roots and associated microorganisms convert sewage to clean water. This new concept is rapidly gaining acceptance because it is the most economical means of treating sewage, especially for rural areas and small cities (Table 1).

Microorganisms have always been used by engineers to treat sewage and industrial wastewater. But the use of higher plants in completing nature's cycle is a new addition to this process. Although microorganisms are a vital part of wastewater treatment, it is important to have vascular plants growing in these treatment filters to feed off the metabolic by-products of microorganisms and to prevent slime layer formation from dead microorganisms. Aquatic plant roots can also add trace levels of oxygen to help maintain aerobic conditions in plant-microbial wastewater treatment filters.

One question often asked is, "Will this wastewater treatment system work in cold climates?" This question has been answered by a small town, Monterey, Virginia. Located in the mountains of western Virginia, near the West Virginia border, Monterey's temperature reaches levels of -30 degrees Fahrenheit. This small town has installed a bulrush/rock filter system to treat their waste. This system has been in operation over two years now and the latest data available indicated it was meeting design treatment levels.

The largest aquatic plant rock filter system

installed to date is at Denham Springs, Louisiana. This system is treating approximately three million gallons per day of domestic sewage. With EPA grant money being phased out, the only affordable alternative for small towns and rural areas is the aquatic plant wastewater treatment system. To demonstrate the effectiveness of aquatic plant wastewater treatment systems, data from a mobile home park in Pearlington, Mississippi designed to treat 10,000 gallons per day is shown (Table 2). The owner has converted his sewage treatment system into a beautiful flower garden containing canna lilies, water iris and elephant ears.

Although the largest number of aquatic plant wastewater treatment systems installed to date have been for treating domestic sewage, the use of these systems for treating industrial chemical wastewater is rapidly increasing (Table 3). The chemical manufacturing industry, paper mills, the textile industry and animal processing plants are beginning to utilize the aquatic plant wastewater treatment process as an economical and environmentally safe method of treating their wastewater (Figure 2). The catfish farmers in Mississippi are also experimenting with aquatic plant filters for treating and recycling their fish culture waters.

Houseplants combined with activated carbon filters are also a promising solution to the complex problems of indoor air pollution. The United States Environmental Protection Agency (EPA) studies have stated that indoor air pollution represents a major portion of the public exposure to air pollution

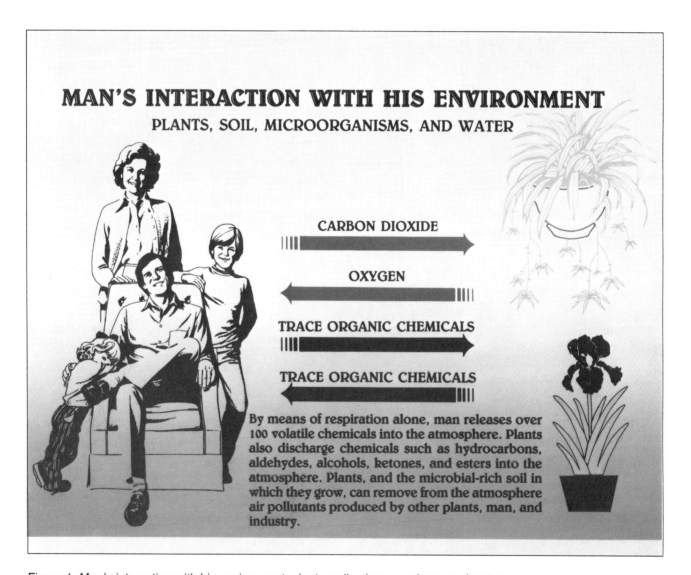

Figure 1. Man's interaction with his environment: plants, soil, microorganisms, and water.

Biological Life Support Systems

SMALL TOWNS, MOBILE HOME PARKS SUBDIVISIONS, AND SINGLE HOMES		INDUSTRY	GOVERNMENT FACILITIES
1.Monterey Va* 2.Albany, La* 3.Benton, La* 4.Crowley, La 5.Choudrant, La 6.Delcambre, La 7.Denham Springs, La* 8.Haughton, La* 9.Livingston Parish,La* 10.Mandeville, La(City)* 11.Mandeville, La 　(Subdivision)* 12.St. Martinville, La 13.Sunset, La. 14.Sibley, La* 15.Collins, Ms*	16.Leakesville, Ms 17.Pearlington, Ms* 18.Pelahatchie, Ms* 19.Union, Ms 20.Utica, Ms* 21.Summit, Ms 22.Picayune, Ms* 23.Terry, Ms 24.Cottonwood, Al* 25.Mauriceville, Tx* (Restaurant & Store)	1.Natchitoches, La* 　(Tenn Gas Pipeline Co.) 2.Theodore, Al* (Degussa 　Chemical Corporation) 3.Columbus, Ms. 　(WeyerhausePaper Mill) 4.New Augusta, Ms (Leaf 　River Forest Products, Paper 　Mill) 5.Sulphur, La (Fredeman Shipyard)	1.Carville, La* 　(U.S.P.H.S. 　Disease Center) 2.NASA, John C. 　Stennis Space 　Center, Ms*

*In operation. All others under construction or in planning and design phase.

Table 1. Aquatic plant wastewater treatment systems using technology developed by B.C. Wolverton, Ph.D..

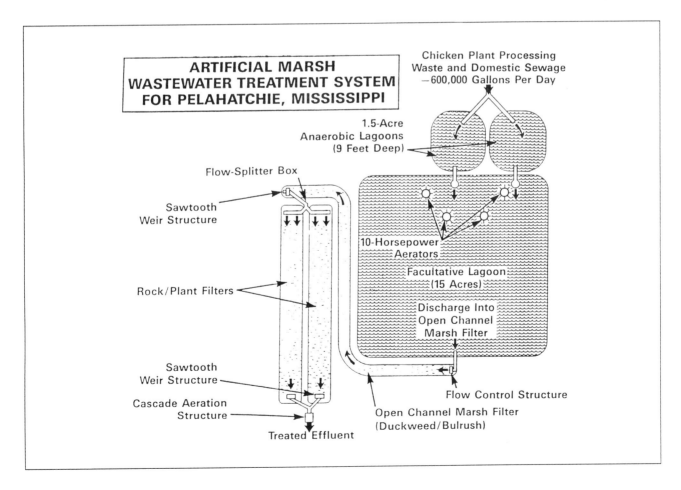

Figure 2. Artificial Marsh Wastewater Treatment System for Pelahatchie, Mississippi.

```
       SEPTIC TANK ROCK/PLANT WASTEWATER TREATMENT SYSTEM FOR
       SUNRISE HAVEN MOBILE HOME PARK, PEARLINGTON, MISSISSIPPI
```

	SEPTIC TANK EFFLUENT (mg/l)			MARSH FILTER EFFLUENT (mg/l)		
Date	BOD$_5$	TSS	Fecal Coliform Colonies/100 ml	BOD$_5$	TSS	Fecal Coliform* Colonies/100 ml
6/88	138	20	5.8×10^5	7.2	0	3,300
7/88	111	16	2.9×10^6	7.5	0	366
8/88	83	0	2.0×10^6	6.5	0	<1
9/88	119	28	1.9×10^6	5.6	0	1,333
10/88	126	136	1.1×10^6	5.2	8	2,000
11/88	95	184	2.2×10^6	2.8	0	1,800
12/88	134	270	3.8×10^6	7.9	8	1,400
1/89	86	24	3.8×10^6	6.6	12	566
2/89	97	46	1.0×10^6	5.4	6	1,500
3/89	86	16	1.0×10^6	6.3	4	3,000
4/89	183	130	1.5×10^6	7.4	5	315
5/89	164	100	6.5×10^6	4.7	0	216

*Before Chlorination

Table 2. Pearlington, Mississippi Septic Tank/Rock Plant Marsh Filter.

ARTIFICIAL MARSHES FOR TREATING INDUSTRIAL WASTEWATER CONTAINING TOXIC CHEMICALS

Chemicals (mg/L)	Marsh Plants in Rock Filter	Influent	Effluent*
Trichloroethylene	Torpedo grass	3.60	0.0009
	Southern bulrush	9.90	0.05
Benzene	Torpedo grass	7.04	1.52
	Southern bulrush	12.00	5.10
	Reed	9.33	0.05
Toluene	Torpedo grass	5.62	1.37
	Southern bulrush	11.47	4.50
	Reed	6.60	0.005
Chlorobenzene	Torpedo grass	4.85	1.54
	Southern bulrush	10.65	4.90
Phenol	Cattail	101.00	17.00
	Reed	104.00	7.00
P-xylene	Reed	4.07	0.14
Pentachlorophenol (PCP)	Torpedo grass	0.85	0.04
Potassium cyanide	Torpedo grass	3.00	<0.20
Potassium ferric cyanide	Torpedo grass	12.60	<0.20

*24-hour retention time

Table 3. Artificial marshes for treating industrial waste-water containing toxic chemicals.

ENERGY-EFFICIENT HOMES WITH BIOREGENERATIVE LIFE-SUPPORT SYSTEMS

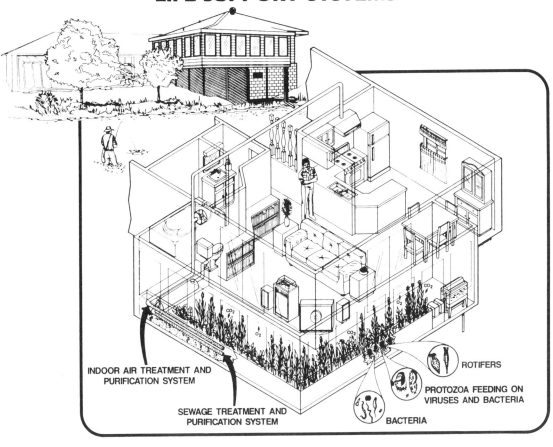

INDOOR AIR TREATMENT AND
PURIFICATION SYSTEM

SEWAGE TREATMENT AND
PURIFICATION SYSTEM

BACTERIA

ROTIFERS

PROTOZOA FEEDING ON
VIRUSES AND BACTERIA

Figure 3. Energy-efficient Homes with bioregenerative life-support systems.

and may pose serious acute and chronic health risks. The EPA studies also state that the potential economic impact of indoor air pollution is estimated to be in the tens of billions of dollars per year.

To enhance the efficiency of common houseplants and potting soil in removing indoor air pollutants, I recently developed a high efficiency plant filter system combining activated carbon and other adsorbent materials into a unique filter system. This patent pending system utilizes a fan to rapidly move polluted air through a mixed bed filter containing a combination of the most effective adsorbent materials in a hydroponic plant growth chamber. The hydroponic reservoir continuously supplies moisture to the plant root zone to prevent the roots from being damaged during continuous operation of the exhaust fan which moves air

through the plant root adsorbent mixture bed. One of the unique components of this process is the utilization of plant roots and microorganisms to continuously clean and bioregenerate the adsorbent bed filter.

The obvious next step in development of plant and microbial filter biotechnology is to incorporate the complex wastewater treatment/indoor air purification concept into a real home environment. This I have recently accomplished in my own home. Although it took some time to convince my wife to allow me to flush raw sewage into a planter system in her house, she reluctantly allowed me to install such a wastewater treatment/indoor air purification system (Figure 3). Now we have a lovely Florida room filled with beautiful houseplants that purify the air while feeding off the wastewater.

With this accomplished, I am now feverishly attacking the air emission problems from smokestacks, incinerators, etc. This is the final part of the puzzle to be completed using green plants and microorganisms for solving Earth's water and air pollution problems. The approach to solving the point source air emission problem is to convert the air pollutants into water pollution and purify the polluted waters using aquatic plant microbial marsh filters.

Since conventional technology has failed to solve the Earth's environmental pollution problems, the most promising option left to man, in my opinion, is to harness the power of nature by using plants and their associated microorganisms to undo man's damage.

ACKNOWLEDGEMENTS

The one person in NASA that supported this effort over the years is Ray Gilbert at NASA's Office of Technology Utilization in Washington, D.C. He believed in this research and supported it with continuous funding. I would like to take this opportunity to personally thank Ray for his support.

REFERENCES

1. Gillette, Becky. 1989. "Artificial Marsh Treats Industrial Wastewater." *Biocycle.* Vol. 30, No.2, pp. 48-50.

2. Marinelli, Janet. 1990. "After The Flush: The Next Generation." *Garbage.* Vol. II, No. 1, pp. 24-35.

3. Wolverton, B.C. 1988. "Aquatic Plants For Wastewater Treatment." *The World & I.* pp. 382-387, December, 1988.

4. Wolverton, B.C. 1989. "Aquatic Plant/Microbial Filters For Treating Septic Tank Effluent," in Constructed Wetlands For Wastewater Treatment, D.A. Hammer, Ed. (Lewis Publishers, Inc., Chelsa, MI), pp. 173-178.

5. Wolverton, B.C., R.C. McDonald, C.C. Myrick, and K.M. Johnson, "Upgrading Septic Tanks Using Microbial/Plant Filters," *J. MS Acad. Sci.* 29:19-25 (1984).

6. Wolverton, B.C. "Hybrid Wastewater Treatment System Using Anaerobic Microorganisms and Reed *(Phragmites communis)*," *Econ. Bot.* 36(4):373-380 (1982).

7. Wolverton, B.C., R.C. McDonald, and W.R. Duffer. "Micro- organisms and Higher Plants for Wastewater Treatment," *J. Environ. Qual.* 12(2):236-242 (1983).

8. Wolverton, B.C. "Artificial Marshes for Wastewater Treatment," in *Aquatic Plants for Wastewater Treatment and Resource Recovery,* K.R. Reddy and W.H. Smith, Eds. (Orlando, FL: Magnolia Publishing, Inc., 1987), pp. 141-152.

9. Wolverton, B.C. "Natural Systems for Wastewater Treatment and Water Reuse for Space and Earthly Applications," In *Proceedings of American Water Works Association Research Foundation, Water Reuse Symposium IV* (Denver, CO:1987), pp 729-741.

10. Wolverton, B.C. "Aquatic Plants for Wastewater Treatment: An Overview," in *Aquatic Plants for Wastewater Treatment and Resource Recovery,* K.R. Reddy and W.H. Smith, Eds. (Orlando, FL: Magnolia Publishing, Inc., 1987), pp. 3-15.

11. Wolverton, B.C., and R.C. McDonald. "Natural Processes for Treatment of Organic Chemical Waste," *Environ. Prof.* 3:99-104 (1981).

12. Wolverton, B.C., and R.C. McDonald-McCaleb, "Biotransformation of Priority Pollutants Using Biofilms and Vascular Plants," *J. MS Acad. Sci.* 31:79-89 (1986).

The BioHome: A Spinoff of Space Technology

Anne Johnson
Microbiologist, NASA Stennis Space Center
Picayune, Mississippi

I would like to preface the discussion of the Bio-Home with some information about the work we have been doing at our environmental laboratory over the past 15 years. The main focus of Stennis Space Center (SSC) is shuttle engine testing, however, we also have a very active laboratory addressing environmental issues related to biological life support.

Some of the earliest work at Stennis pioneered the utilization of water hyacinths for wastewater purification. This technology has been utilized throughout the world for treatment of both domestic and industrial waste.

The SSC environmental laboratory has also done a lot of research in the field of artificial marshes. These systems are essentially 18 inch deep pits in which 30 mil plastic liners have been installed. The rock substrate is then added to support various types of vascular plants such as canna lilies or bulrush. Artificial marshes are very effective in terms of reducing fecal counts in the effluent. Surprisingly, there is typically no odor associated with such a system. These factors, along with the system's low cost and esthetic quality have made it very appealing (Figure 1).

One of the primary goals of our laboratory is technology utilization. Simply this means that as we develop and refine technology, it is also our role to provide this information to the public as well as to aid in its implementation. Certainly the wastewater systems previously mentioned are such an example.

It should be noted that all of the wastewater at SSC is treated by these types of systems. We hope to be able to use this technology in conjunction with Space Biosphere Ventures to evaluate these systems in Biosphere 2.

Five or six years ago, we began looking at the problem of indoor air pollution. Many people are familiar with the problem of formaldehyde contamination. This chemical is known to leach from formaldehyde resins used inside buildings. However, there are a variety of other potential pollutants that you may encounter in an indoor environment including benzene and trichlorethylene, common constituents of paints and solvents.

We became interested in evaluating a biological system comprised of plants and microorganisms for the purpose of reducing organic contaminants. Initial studies involved placing various plants in plexiglass chambers and injecting known quantities of pollutants. The changes in concentration were measured by gas chromatography. We have since expanded the types of plants screened as well as the number of pollutants involved. We are also interested in the possible synergistic effects that may be occurring when organic substances interact. Also, we will be addressing the possible fluctuations that may occur with respect to pollutant concentration.

We have just completed a two year joint project with the Associated Landscape Contractors of America (ALCA), where we screened several foliage plants for their ability to reduce concentrations of benzene and trichloroethylene. Of the two, benzene is most easily reduced. However, plants such as the Chinese evergreen, peace lily and mother-in-law's tongue exhibited the capability to reduce concentrations of either pollutant. At this point, we are interested in figuring out what the mechanism behind the purification scheme is. We feel it is a symbiotic relationship between the plant roots and the associated microflora. A preliminary microbial profile indicates that the required microorganisms are common soil types. Future plans call for further microbiological analyses as well as exposure of the plants to radio-labeled pollutants. The latter will enable us to ascertain the regions of the system where the pollutant resides.

The plant filter has gone through several design changes. These are constructed on site with materials readily available off the shelf from stores like K-Mart. We started out using plants in a potting soil/lava rock/charcoal substrate, followed by soil and charcoal alone. The present filter incorporates a fan system which functions to pull room air across the soil/charcoal interface (Figure 2).

One of the main concerns that we have, especially with respect to a closed environment, is whether or not these systems are expelling microorganisms into the air. We are presently conducting analyses to determine the numbers and types of microbes that are emitted.

The BioHome is a 650 square foot habitat that will enable us to evaluate the efficiency of bio-regenerative technology in a closed system. The structure is 46 feet long and 16 feet wide with 12 inch thick fiberglass insulation. This facilitates maintenance of indoor temperatures over a narrow range. Although there is restricted air flow, the system is not closed at this time. However, the air conditioners are designed such that they do not introduce outside air, but rather recycle that from the interior (Figure 3).

The BioHome is divided into two areas: the living area and the waste treatment area. In the

Figure 1. Artificial marsh for wastewater treatment.

Biological Life Support Systems

former, plant filters have been included in an effort to dissipate off-gassing products. Either the plastic mats on the floor or the laminate from the wall tiles release a substance which tends to irritate the eyes and respiratory tract. (Figures 4 and 5)

The wastewater facility is essentially a small artificial wetland system adapted for inclusion in the BioHome. Wastewater flows from the exterior septic tank into a series of 8 inch diameter PVC pipes and finally into a 100 gallon aquarium. Segment 1 of the pipe is empty in order to facilitate further settling of solids. Segments 2 and 3 are approximately 50% full of lava rock which functions to promote development of a biofilm. Plants such as canna lilies and water iris are also included. Segments 4 and 5 also have plants but the substrate inside is granular activated carbon, while segment 6 includes carbon in the first few feet of pipe, followed by a substance known as zeolite.

The latter functions to remove ammonia from the system. (Figures 6 and 7)

We are also in the process of growing numerous types of vegetables such as peas, tomatoes, and cabbage. As plants from the wastewater system die, they are removed and used as compost for the vegetable plants.

There is also a system utilizing plants to provide a source of drinking water. As water vapor is produced by the plants via evapotranspiration, a dehumidifier removes water from the air. From this point it is filtered by means of an activated carbon substrate, then treated with ultraviolet light prior to collection.

The living quarters of the BioHome comprises approximately one half of the total square footage. We have had an individual occupy the structure for a period of several months. Their primary function was to provide a source of waste for the system.

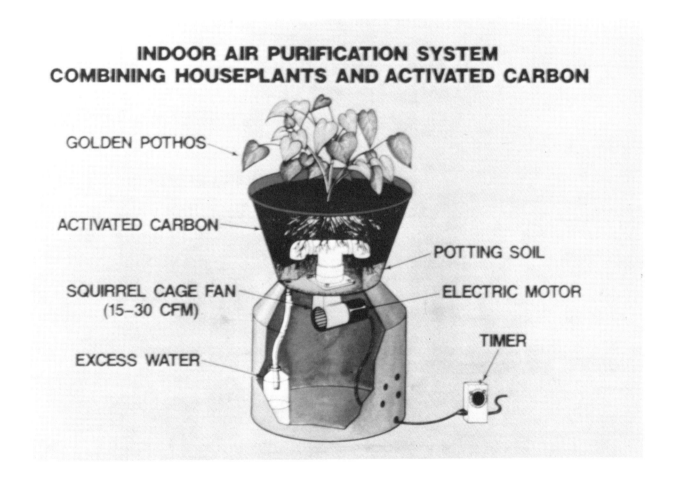

Figure 2. Indoor air purification system combining houseplants and activated carbon.

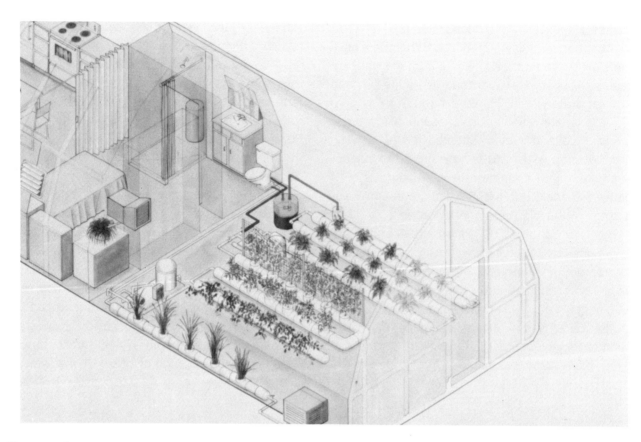

Figure 3. Cutaway section: the BioHome.

Figure 4. Interior of BioHome prototype: dining/study area.

Biological Life Support Systems

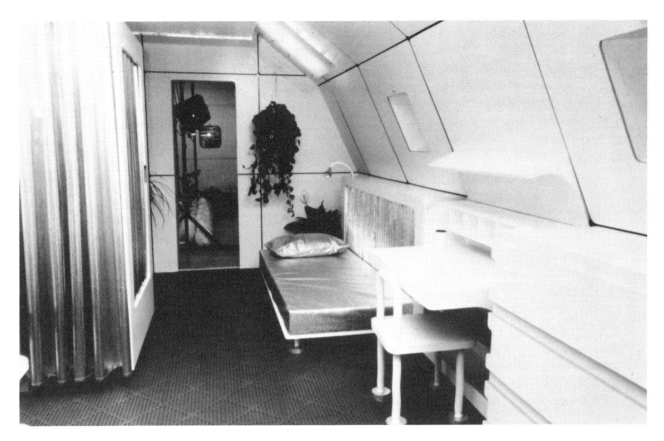

Figure 5. Interior of BioHome prototype: bedroom area.

Figure 6. Miniaturized artificial marsh wastewater treatment system for the BioHome prototype.

At this point, the only recycling of water that occurs is the pumping of treated water into the toilet. The majority of reclaimed water flows into a 100 gallon aquarium which serves as a buffer. In the event that the water level drops below a certain point in segment one, a sensor will function to turn on pumps in both the septic tank and aquarium to replenish the system.

We are screening several types of plants for inclusion in the wastewater facility including torpedo grass and canna lilies. It is important to keep in mind that different locations in the system have different organic concentrations. Therefore we are screening various plants to determine which will fare best in regions with a high organic content as opposed to those regions that do not.

We are also looking at inclusion of halophytic (salt-tolerant) plants into the system. This could be a useful addition since human waste is typically high in salt.

As I mentioned previously, the source of drink-ing water is water vapor obtained from plants. There are so many plants in the treatment facility at this point, it bears resemblance to a jungle. However, only about 11 liters of water are produced per day. This is not enough to accommodate one person's drinking water, bath and cooking requirements. Consequently our goal is to increase this volume of water production by adding additional plants.

We do an extensive array of tests on water from the BioHome. Typically, water quality results fall well within the necessary guidelines. We are also concerned with the possible presence of volatile organics in the air and consequently have instituted a sampling regimen.

The next few months will be devoted to extensive biological and chemical analysis in order to determine what types of microorganisms and chemicals may be found within such a system. With this information in hand, studies will expand to incorporate the presence of humans.

Figure 7. Detail of BioHome artificial marsh wastewater treatment system.

Earth Observing Satellite:
Understanding the Earth as a System

Gerald Soffen, Ph.D.
Associate Director for Program Planning and
Chief Scientist, Earth Observing Satellite
Goddard Space Flight Center

INTRODUCTORY COMMENTS

The Spirit of Discovery

While my topic is the biosphere of the Earth — Biosphere 2's big brother — my intellectual life really started with Mars. I have thought about what would be the number of a biosphere on Mars? It can't be Biosphere 3, and not infinity, because something that supported life in another solar system might be infinity. But Mars is certainly the most likely planet we know that could become a biosphere.

I will relate a story about the Viking Mission to Mars. I had a very rich experience here listening to the papers and asking questions. I am going to start by telling you that I was terribly jealous of what I saw today. I had the same impression that I had in 1961 when I first joined NASA — and that is the youthful spirit of enthusiasm that is easily recognized. It was really an exhilarating experience to suddenly find, "It feels around Biosphere 2 like it was in 1961 when the space program was just starting." One didn't care about making "mistakes". You couldn't make mistakes, you didn't even conceive of making mistakes — you went ahead and did things. Last night, Margret Augustine told me about the Institute of Ecotechnics' ship, the *R/V Heraclitus* that was raised after Hurricane Hugo sank it in San Juan harbor in September 1989. It was an extraordinary story of quick response and ingenuity in the rescue. I am afraid NASA couldn't do that today. If they did, they would do it in spite of the system, not because of the system. In a sense, you didn't "know any better" so you just went ahead and did it. That's the story of what's happening at the Biosphere 2 project and why I and your other visitors are so in love with what they are seeing — Biosphere 2 recaptures that sensation of discovery and exploration.

My Viking story starts with the great ocean explorer Jacques Cousteau. The Viking Project was ordered to land on Mars on the 4th of July 1976, the 200th anniversary of the birth of the United States. As it happened, that date was exactly in the right window, the right several weeks during which Viking could have landed. So, of course, we planned the landing for the evening of July 4. Unfortunately, Mars didn't behave. After the spacecraft arrived at Mars on the 20th of June, we took one look at the planetary surface and I said, "There goes the ball game." There was no way it could get down on the landing site that had been selected earlier and land safely. That site was in the midst of an area of extremely steep mountains and canyons.

We suddenly realized that we had a problem on our hands. We were going to have to find a new landing site. As you know, we did find a landing site and landed two spacecraft successfully. But on the evening of June 22 we had just made the decision not to attempt a landing on the 4th of July. The project manager turned to me and said, "What are we going to do for the 4th of July. We have four hundred people from the press that are showing up at Jet Propulsion Lab. They are going to be covering something, and if we are not landing, what are we doing?" CBS, NBC, ABC all had their crews ready, they were moving in, the big TV and film trailers were already parked there. So we dreamed up the idea of a symposium on the concept of "discovery" with a really distinguished group of people as speakers — Carl Sagan, Ray Bradbury, Norman Cousins, editor of *Saturday Review,* Jacques Cousteau and Phil Abelson of *Scientific American*. Each gave wonderful and stimulating talks.

Afterwards I had supper with Jacques Cousteau, and he talked about what he does on his research ship *Calypso*. He said most of the trained scientists aboard have a vertical view. Their job is as experts and specialists. They go into the water to investigate these specialties and every question they ask is a more profound one. Cousteau said, "I am a horizontal scanner. I am not an expert in anything. I am the person who keeps my eyes open, watches from the sides and be-

comes aware of where things might come together that otherwise might not come together."

Discovery doesn't always come from vertical sounding, it also comes from horizontal perspective. As the project scientist, my job on Viking was sort of as a symphony director — you don't play anything but you are supposed to keep everybody else playing the right music. We had seventy scientific investigators on Viking. I said to Cousteau, "I loved your story, how can I ever develop this great vision that you have and ability to observe the total picture from the side." He replied, "I don't know if you have any talent in this area, but you seem to have the interest in it, so that insight will carry you a long way."

Viking was a great success, but the end of my story is this. From that evening when we first viewed the Martian surface till the spacecraft landed on July 20, a most intensive activity of

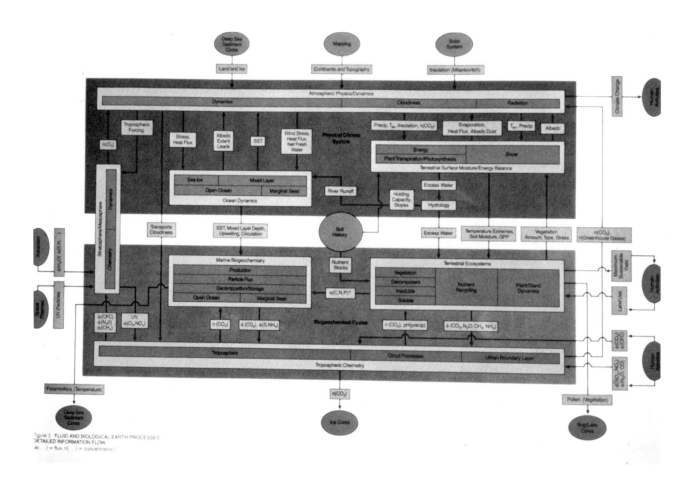

Figure 1. Fluid and biological Earth processes, detailed information flow chart. (NASA)

Biological Life Support Systems

landing site selection involved not just planetary experts, but a whole wealth of remarkable talent, sitting as we were on the doorsteps of Cal Tech.

The answer was interesting for it came from a scientist no one expected: a geochemist from Princeton who had a tiny little experiment using a magnet on the end of the Viking digging arm. Dr. Rob Hargraves' experiment was simple: to see what sticks to the magnet. Hargraves is very quiet and very smart. He had breakfast with me one day and said, "We are struggling so hard to land our spacecraft on Mars. We are counting craters, we are trying to reconstruct the history of Mars, we are looking at all the signs, we are looking at Mariner data, we are asking the Soviets...how about just finding out which way the wind blows and try to land where the soft spot is?"

It was so obvious! Every high school student could have thought of that. I said, "Have you asked the meteorologist that?" He said, "It just occurred to me." And that eventually led to the solution of the Viking landing site selection process. I was in the right place at the right time to recognize a good idea and Rob Hargraves had the right idea.

After Viking was over, I decided that as a biologist, I was dealing with the wrong planet. As a biologist, going to Mars with life systems would be wonderful; but Mars is not the most promising place for exploration for a biologist. Its very likely from what we now know that there is no organic material on Mars. When I finished Viking, I got interested in the Earth. Recalling that marvelous experience we'd had during those intense days of looking for a landing site, I wondered about a comparable situation that affects the study of Earth. The Earth is partitioned — there are oceanographers, meteorologists, chemists, agronomists, and so on, who rarely talk to each other. When we were trying to land on Mars, everybody talked to everybody. Whether you were a meteorologist, geophysicist, you only cared that the project worked, the project was everything. That's what I see happening here at the Biosphere 2 project. It doesn't matter if you're an agronomist, engineer, entomologist, metallur-

Eos Baseline Planning Scenario

NASA POLAR PLATFORM-1 NPOP-1	NASA POLAR PLATFORM-2 NPOP-2	ESA POLAR PLATFORM-1 EPOP-1/A1	ESA POLAR PLATFORM-2 EPOP-2/B1	JAPANESE POLAR PLATFORM JPOP	NOAA FREE-FLYER	ATTACHED PAYLOADS
Orbit: 705 km Crossing: 1:30 pm Launch: 4th Qtr. 1996	Orbit: 705 km Crossing: 1:30 pm Launch: 4th Qtr. 1998	Orbit: 824 km Crossing: 10:00-10:30 am Launch: 1997	Orbit: 705 km Crossing: 10:00-10:30 am Launch: 2000	Orbit: 800 km Launch: 4th Qtr. 1998	Orbit: 824 km Crossing: 1:30 pm Launch: 1st Qtr. 1998	Orbit: 400 km Crossing: 28.5° Launch: 1st Qtr. 1998
AIRS	SAR	ATLID	ATLID	LAWS	ARGOS	MODIS-N
ALT	SEM	MERIS	HRIS	AMSR	AMRIR	MODIS-T
GLRS	GGI	MIMR	MIMR	AVNIR	AMSU	MIMR
HIRIS	GOS	AMIR	SAR-C	OCTS	GOMR	AMSR-2
MODIS-N	IPEI	CHEMISTRY	AMIR	SAR-L	SEARCH & RESCUE	ERBI
MODIS-T	LIS	RADIOMETER	CHEMISTRY	SAR-X	SEM	MIMR
SEM	MLS	ALTIMETER (GPS)	STEREO IMAGER			OZONE SENSOR
MIMR	SAFIRE*	AMI-2				PPS-PODS
AMSR	SWIRLS	AMRIR				RAIN RADAR
ITIR (TIGER)	TES	AMSU				SCATTEROMETER
CERES	XIE	SEM				SPECIAL IMAGER
DLS		ARGOS				
ENAC		SEARCH & RESCUE				SOLAR FLIGHT OF OPPORTUNITY 1st QUARTER 1995
EOSP						
GGI						ACRIM
HIMSS						SOLSTICE
HIRRLS						
IPEI						
MISR				NASA RESEARCH FACILITY INSTRUMENT		
MOPITT				ESA RESEARCH FACILITY INSTRUMENT		
POEMS				JAPANESE RESEARCH FACILITY INSTRUMENT		
SAGE III				OPERATIONAL FACILITY INSTRUMENT		
SCANSCAT				PI INSTRUMENT INVESTIGATIONS		
TRACER						
AMSU						

Figure 2. Earth Observing Satellite (EOS) Baseline Planning scenario. (NASA)

gist, ecologist or architect — the project is driving you together and making you think how one element affects the whole system. That spirit is what we have to achieve to preserve the Earth.

EARTH OBSERVING SATELLITE

From one perspective, the Earth Observing Satellite (EOS) came out of what we did on Mars, out of that struggle for a landing site. When I left the Mars project and reentered the "real world" at NASA Headquarters, I took a job as Director of Life Sciences. I met Dan Botkin of U.C. Santa Barbara, one of the participants at this workshop. At the time he was an advisor for the National Academy of Sciences Space Biology Board and was writing a

report on life support systems. The report talked about closed ecological life support systems and in conversations with Dan I began realizing that the ultimate life support system we know is the Earth, the global biosphere. From that came the realization that if we are ever going to do anything we had to start an effort in global ecology. We started with a program that was the predecessor to Dr. Mel Averner's program in biospherics. It was called Global Habitability — then in NASA parlance "System Z" — and has changed its names over the years but the basic concept has always been to study the Earth as a planet (Figure 1). This means to study the components of the Earth as they fit together — as you are doing in Biosphere 2 — not to go our separate ways. This is what I think hu-

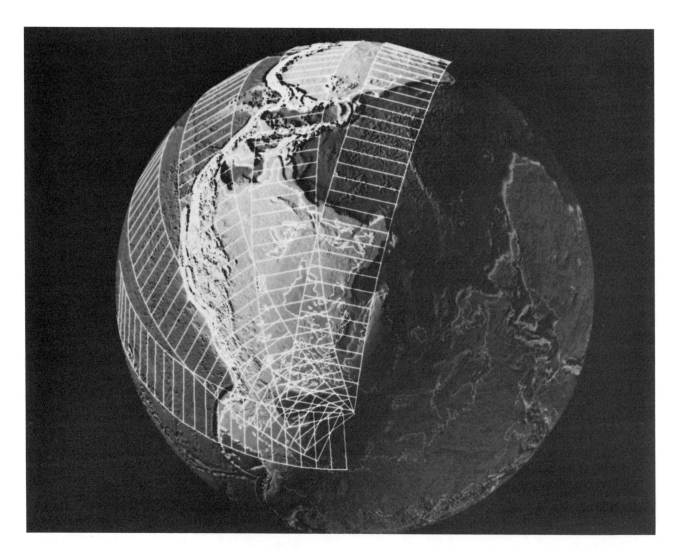

Figure 3. Earth-mapping from a polar orbiting satellite utilizes the Earth's rotation for planetary coverage. (NASA)

mankind is going to have to do if we are going to have any biological survival in the next century. We have a problem trying to communicate with and educate one another. It is almost like the Tower of Babel trying to talk together. Even many of those scientists who want to do it don't yet understand a common language, but we are learning how.

About five years ago, a group of scientists began to coin the name "Earth Systems Science". They began looking at the Earth the way an engineer might look at it, as a whole system. This is the way you look at Biosphere 2 — you see the whole system! Today we toured the heating system, the electrical system, the biological systems, the control system, etc. Similarly, we began to see how all of the Earth pieces are linked. Shortly after we started, another effort began. Partly out of alarm, partly as a result of concern about acid rain, of other global changes that were occurring, there was an attempt on a grand scale to respond.

NASA, because it is a responsive agency, volunteered. We didn't know any better so we just proceeded. Once we began to get a little attention, NOAA (National Oceanic and Atmospheric Agency) and EPA got into the act, and the next thing we had all of our political brethren down our necks saying, "Hey, you are doing our stuff. You shouldn't be doing oceanography because NOAA does oceanography, and you shouldn't be doing atmospheric studies because EPA does atmosphere." Right now the Department of Energy (DOE) is concerned because they want to do the carbon dioxide studies, since it comes from the burning of coal and oil which are energy sources. But the important issue is really not who does it, but that these important studies be undertaken.

There is now a plan for global studies which include two very large efforts. One is the International Geosphere/Biosphere Program (IGBP) sponsored by the International Council of Scientific

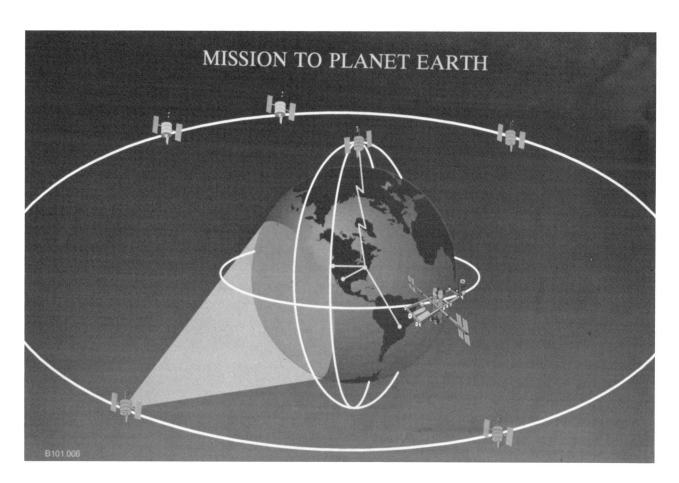

Figure 4. Schematic showing the proposed Mission to Planet Earth satellites. (NASA)

Unions. The IGBP is being discussed and planned in many countries now and is motivated by true concern and alarm about the relevant issues.

The other initiative is Mission to Planet Earth, an umbrella program for doing three kinds of space missions. The major one is the Earth Observation Satellite for which I am Project Scientist. I count this as a rare privilege, as it is rare even to get a chance to do this once. So having been Project Scientist for Viking, here I get a second opportunity. In addition to EOS, there are two complementary NASA space missions. These involve satellites in sequential orbit as a companion of *Freedom* Space Station and small satellites in geosynchronous orbit. EOS is large polar orbiting satellites with payloads weighing several thousand kilograms

and a total weight of 9000 kilograms. Two will be placed in polar orbit by NASA, one by the Japanese and one or two by ESA (European Space Agency) (Figure 2).

The beauty of a polar orbit is that by observing from pole to pole the Earth turns underneath the satellite and you get to see the entire globe (Figure 3). It lets the Earth do the work as the spacecraft orbits, obtaining fifteen passes a day of the Earth by this EOS containing about a dozen remote sensing instruments. The instruments will be different on each of the polar orbiting EOS as we are wanting to utilize some thirty instruments during the program to have a fairly complete range of sensors mapping the Earth.

There are several key points to the potential

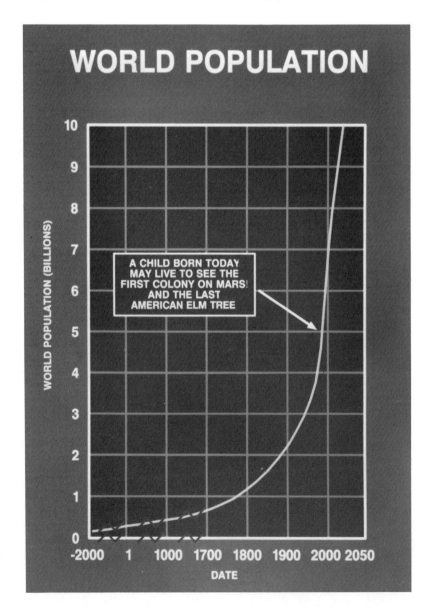

Figure 5. World Population Chart, 2000 BC to present, with human population projections to 2020 AD. (NASA)

significance of EOS. One is that EOS is not focussed on hardware, but on obtaining understanding of Earth processes. It makes a big difference if you are simply trying to launch hardware and obtain data, or whether you're trying to gain understanding. We know that we have to learn how to predict the changes occurring on the Earth. That is the objective — not simply getting more or better data. We have to learn what is causing global change. The test of understanding is prediction. For instance, can we learn to accurately predict global warming trends?

Another key issue is that there are no quick answers to these issues. Most can be answered only by statistical approaches, which require long-term data bases. The changes are subtle enough, and embedded in cycles of varying time-periods, that we must think about placing satellites in these polar orbits not for a quick look, but rather for periods of about a decade. NASA is not used to

doing that, but we are building this requirement into the program to ensure a continuous record of measurements. Why a decade? Basically that is because it is about a solar cycle of eleven years. We may eventually have to look at a longer period, but we know we get sufficient oscillations over a timeframe of one to two years that we must have a longer period. Although many people use the term "solar constant", we in fact know that it is not constant, so at least one solar cycle should be studied. This raises very interesting questions of reliability because no one has previously built spacecraft to last that long. We are currently examining two options. One is to provide servicing capability in space by replacing filters, lenses, batteries etc. by robotic or astronaut flights. The other is to send a replacement craft after 4-5 years in case the first stops operating. Our concerns are that we may spend too much money making the

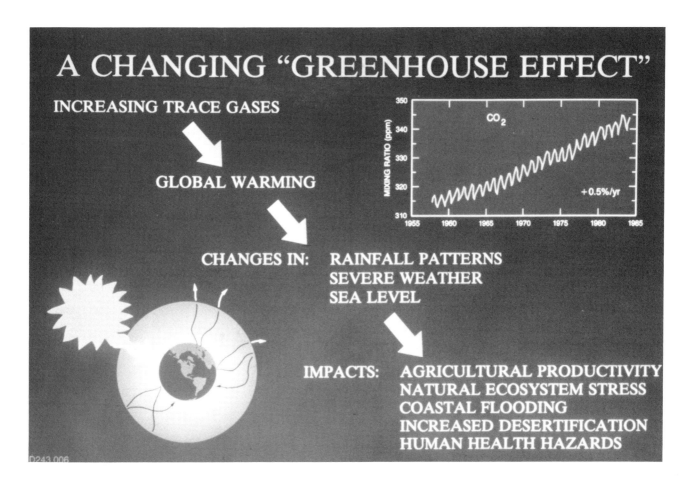

Figure 6. Carbon dioxide levels in Earth's atmosphere 1957-1985, and possible impacts on global environment. (NASA)

payload so reliable that it lasts ten years while it may become obsolete during that time-period.

Another innovative approach of the EOS program is that it is no longer adequate to simply supply the data obtained to one scientist to determine the answers. It is going to be everybody's data. Previously it was given to one scientist, the Principal Investigator, and he was expected to publish his results in due course. Once the EOS data is available, we must get it out, available to anyone who is willing to study it. It is so vital we can't stop with distributing it to just one scientist. The data belongs not only to the United States, it belongs to the world. That makes the EOS Mission different from anything we have ever done.

Another thing that makes EOS unusual is that

it is expensive, very expensive. The requirement of the decade-long operation is a factor that drives up the costs. Recently we calculated the price tag of EOS during various phases. In NASA terminology, Phase A is when you conduct a study of what needs to be done and various options are considered. Phase B is the organizational stage when you select scientists and make determinations of instrumentation. This is the phase we are currently in the midst of. The next phase, Phase C starts about a year from now, in the fall/winter of 1990. This is when expenditures really escalate. The money starts pouring out because we "cut metal", we actually "bend hardware", we really start making things. We are currently planning on spending around 1.2 billion dollars a year. That is an enor-

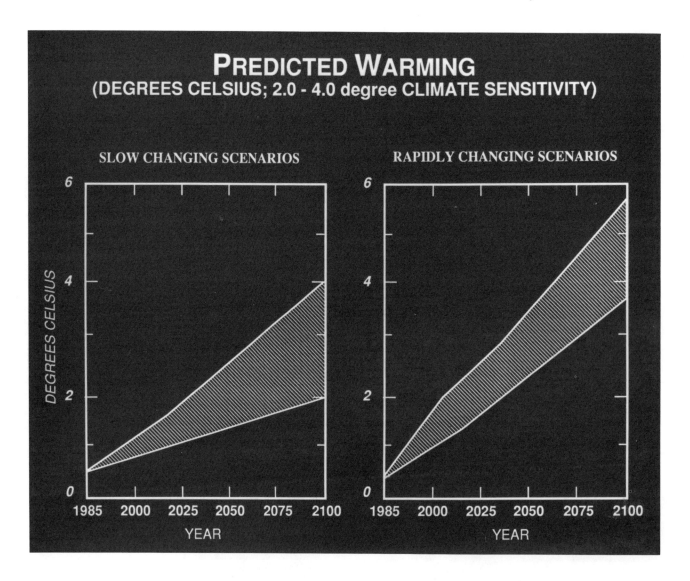

Figure 7. Predicted global warming, showing scenarios of slow and rapid change. (NASA)

mous amount of money. There will be tens of thousands of people working on this effort including engineers in many of the aerospace companies. We will operate at that level of expenditure until we launch the first EOS at the end of 1997. The second one is scheduled for launch two years later.

In conjunction with the American efforts, the Japanese and the Europeans have each committed to do one Earth Observing Satellite also. So four spacecraft — one European, one Japanese and two American — will all be up in polar orbit at the same time. Some of our instruments will be on the Japanese birds and some of the Japanese or the European instruments will be on our birds. The effort will no longer be just ours or theirs — rather, a joint effort in which nations of the world have shared this effort and the data is going to be available to everybody. That's the heart of what EOS is about.

We can summarize the overall mission measurement objectives of EOS:

1. The global distribution of energy input to and energy output from the Earth.

2. The structure, state variables, composition and dynamics of the atmosphere from the ground to the mesopause.

3. The physical and biological structure, state, composition and dynamics of the land surface, including terrestrial and inland water ecosystems.

4. The rates, important sources and sinks, and

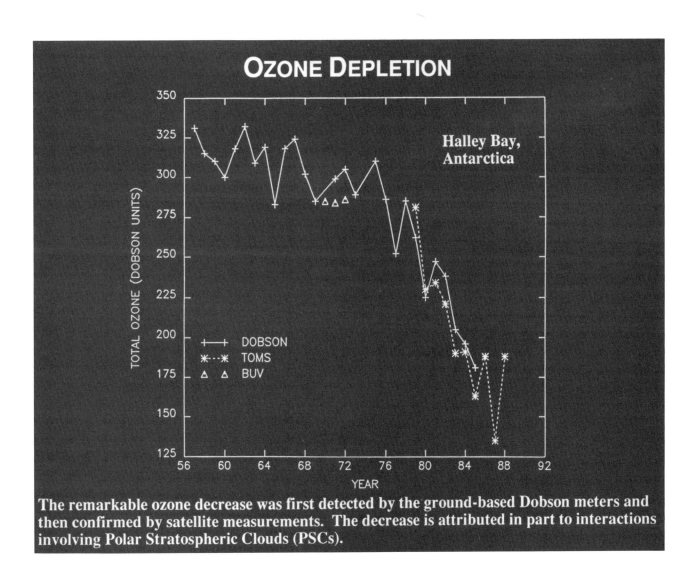

The remarkable ozone decrease was first detected by the ground-based Dobson meters and then confirmed by satellite measurements. The decrease is attributed in part to interactions involving Polar Stratospheric Clouds (PSCs).

Figure 8. Ozone depletion from 1956 to present. (NASA)

key components and processes of the Earth's biogeochemical cycles.

5. The circulation, surface temperature, wind stress, and sea state, and the biological activity of the oceans.

6. The extent, type, state, elevation, roughness and dynamics of glaciers, ice sheets, snow and sea ice and the liquid equivalent of snow in the global cryosphere.

7. The global rates, amounts and distribution of precipitation.

8. The dynamic motions of the Earth (geophysics) as a whole, including both rotational dynamics and the kinematic motions of the tectonic plates.

Earth: The Overview from Space

The space program gave us our first views of Earth as the blue planet and unique life oasis. Those pictures have had an enormous impact — as evidenced by how often they're used. It shows you the impact of such overview pictures. Figure 4 is a schematic showing the overview of Mission to Planet Earth satellites. In conjunction with our polar orbits, there are satellites planned for geostationary orbits, that is missions that are placed in an orbit that allows continuous monitoring over a particular region of the world. If there is an enormous forest fire or volcano that erupts, we have to keep looking at the same place and for this rather than a polar orbit a stationary orbit is required. The problem with geostationary orbits is that they are higher and more expensive to launch. In order to get a station-

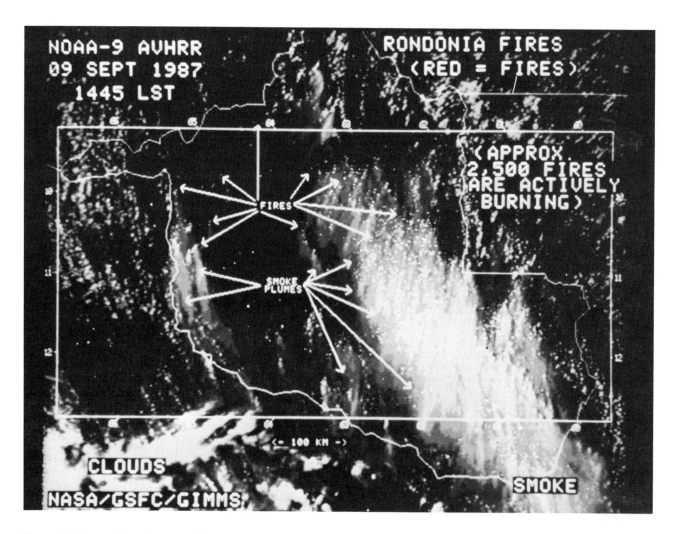

Figure 9. Space view of state of Rondonia, Brazil showing fires from destruction of tropical rainforest. (NASA)

ary orbit the satellite has to be 23 thousand miles out from the surface of the Earth.

Since the Space Station *Freedom* wil be in equatorial orbit, we are going to attach some instruments to it to look at the tropical belt. The tropics are of particular concern because of the enormous deforestation of the tropical rainforests and desertification of tropical grasslands. Underlying the urgency of programs like EOS is the realization that global change is inevitable and of a different character than ever before. Our biosphere seems to be in stress.

The Earth has always been subject to change from what we know of its geological and life history. What is different about the present time is that humans now are beginning to add their own impacts to the natural changes because there are so many of us. We are currently some 5.3 billion people on the Earth. Projections show that this could increase to some 10 million by 2025 and 14 million by the end of the next century (Figure 5). That is an incredibly powerful vector driving many types of other impacts, such as pressure on natural resources, urbanization, pollution etc. and one which can't be simply slowed down.

I want to underline that EOS is not fundamentally hardware, it is the information system to enable us to understand the Earth system. Until recently, we probably didn't have the capability of doing anything like this mission. We didn't have the

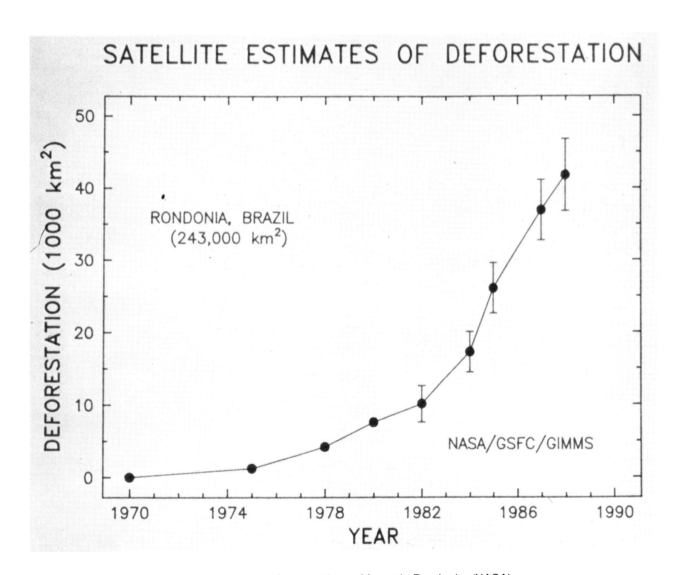

Figure 10. Estimate from space observation of percent loss of forest in Rondonia. (NASA)

computer power we have now. We didn't have all of the technical tools we have now. But we now have no excuse; if we don't do this, or somebody else doesn't do it, it is just because we lack the resolve or are lazy or ignorant or afraid. We certainly have the technical capability now. We must also develop an increasingly interdisciplinary approach to go from data to information to understanding.

We can presently identify pretty obvious major Earth problems. We might call them our generation's "Four Horsemen of the Apocalypse" — acid rain, ozone depletion, global warming and deforestation. These are fueled by the unprecedented rise in human population, and if current trends persist, we haven't seen anything yet. Yet

we don't know other problems that may be coming. These four are just 1989's problems. What may happen in the next five years even if we are able to go ahead with the preparation for EOS when we discover there is a fifth one? Its clear that we are going to have to be able to adequately respond in real time to the unexpected.

Figure 6 shows a graph of increasing carbon dioxide concentrations in the atmosphere and some of the expected consequences. It is perhaps the classical graph that marked our first recognition that there was a very serious problem on Earth. This rise in carbon dioxide is from 1955 to 1985, just a 30 year period. This was originally started by one young man, David Keeling, who went to the top of Mauna Kea in Hawaii and put up a small

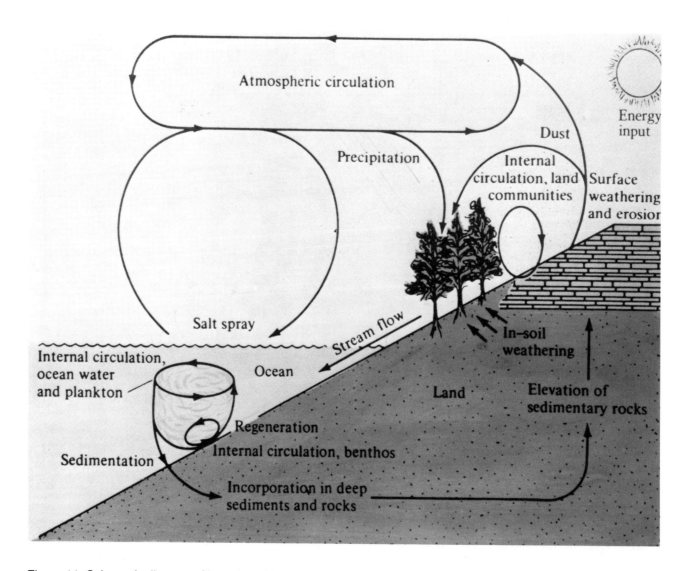

Figure 11. Schematic diagram of interplay of forces determining Earth processes.

sensor and started measuring carbon dioxide in the atmosphere. People thought, when he started reporting his data in the early 1960's, "So what, it doesn't really mean anything. The small amount of carbon dioxide in Earth's atmosphere goes up and down every year, depending on the seasons. It can't be too important." By the late 70's scientists began to realize that the observed rise of about a half of a percent a year could be very serious indeed. I don't want to overemphasize or underemphasize it, we really don't know yet how serious this rise could be. We just don't know the consequences.

Two scenarios have been advanced and modeled by scientists on possible global warming due to the greenhouse effect caused by increasing carbon dioxide in the atmosphere. One is called slow changes and the other is called rapid changes (Figure 7). The latter is a more pessimistic view in which we see a 4-6 degrees C. rise versus the former which results in a 2-4 degrees C. rise in

average temperature over the next century. These increases may not sound like very much, but the last ice age was only 2 degrees C. average temperature colder than our present level. The difference in temperature between where we are now and where we will be in just a hundred years in the most optimistic point of view is about the same as the rise in temperature from the last ice age to now. Our global temperature hasn't changed very much in that long stretch of time. In the most pessimistic view, we could see a 6 degrees rise — or three times as much change. This could have a quite dramatic effect on our climate: the whole western United States is in trouble, north Africa, northern Australia, southwest Asia etc. It is uncertain that this will occur, but it is a least one real possibility that our modelers have worked with.

Figure 8 deals with ozone depletion. We keep hearing about the hole in the ozone layer. So what? What does it mean? The ozone layer protects the biosphere from biologically very potentially damag-

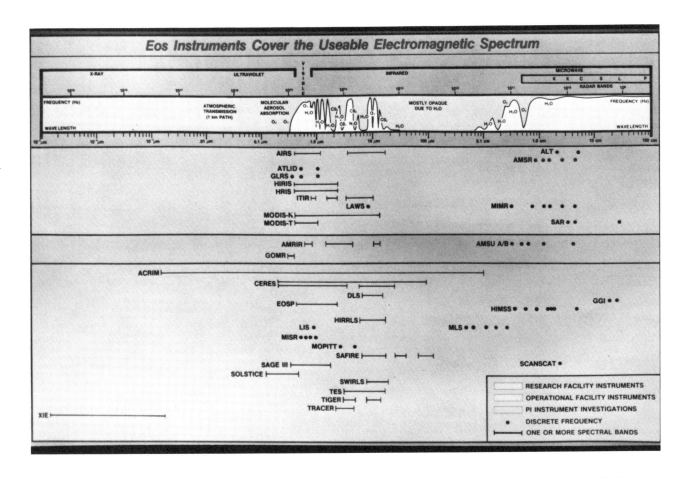

Figure 12. Schematic showing coverage of the electromagnetic spectrum by proposed EOS instruments. (NASA)

ing ultraviolet rays from solar radiation. The depletion of ozone on the Earth from the levels measured in 1956 until the present are quite dramatic. This process is continuing to happen and that is what has us so upset. It hasn't stopped and we don't fully understand it. We do know it is urgent that we find out what is happening to the ozone layer.

If you were an astronaut and flew in the last Shuttle and passed over Brazil, you could look down and see Rondonia, which is a state about the size of Arizona. You would see quite clearly a gridwork of roads where they are cutting the tropical rainforest down. In Figure 9 we have a space view of the state of Rondonia, showing that about a quarter of the state of Rondonia is completely smoke-ridden with approximately 2,500 fires burning at the same time. If you are on the ground it will look like a scene of great devastation. In fact, the deforestation of Rondonia has currently resulted in a loss of tropical forest from about 40 percent of the total area. Figure 10 graphs the rate at which this deforestation is happening.

Scientists who have been studying this realize they must start putting together models to try to understand some of these problems. We must understand where the Sun plays a role, where the clouds play a role, what does the ocean do, what does the land do, what does the water do? All of that has to be put together. We have never tried to understand this whole Earth as a planet. This EOS mission has many different objectives that cover so many areas dealing with water, energy, biology and

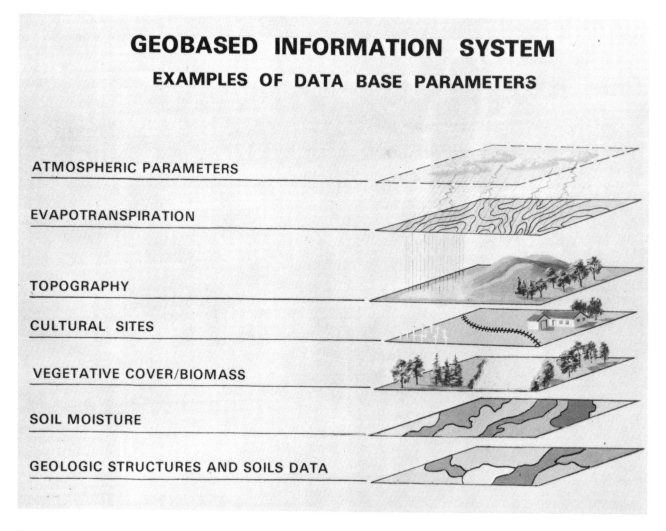

Figure 13. Geobased information system: examples of database parameters. (NASA)

ocean systems. In fact, there are about 500 different things we are going to try to measure dealing with such elements as winds and clouds and volcanoes and sea surface. One thing that makes the Earth so complicated, so much harder to understand than Mars, is that we have an enormous amount of liquid water that changes the whole character of the planet. By comparison, Mars is bone-dry, though it may have large amounts of frozen water below its surface, compared to the Earth. On top of that are the myriad effects of life and additionally the impacts and activities of humans. This planet is so complicated that we are beginning to realize how hard this task of understanding will be (see for example, Figure 11).

Remote sensing instruments that we use have to look down through a complex, layered and often clouded atmosphere. Figure 12 shows a schematic of the electromagnetic spectrum that will be covered by EOS instruments. Both in the visible range

and the microwave range, there are a variety of instruments planned to look down through the atmosphere. Each of the EOS spacecraft have different combinations of instruments to do different jobs and together they mesh together to try to get the total understanding. Each of these instruments is the size of a large wardrobe — much larger than instruments we have flown previously in spacecraft. We need large instruments because the measurements are complicated. One of the problems of the polar mapping process is that it would leave holes in the equatorial region. That is the reason that we want the equatorial mission associated with Space Station *Freedom* to complement EOS.

Figure 13 gives an illustration of the data base parameters. We have to take all of these various pieces and fit it all together. We will receive about a trillion bits of data every day for ten years during EOS. In the language of computer scientists that is a terabyte of data per day. A terabyte of data is

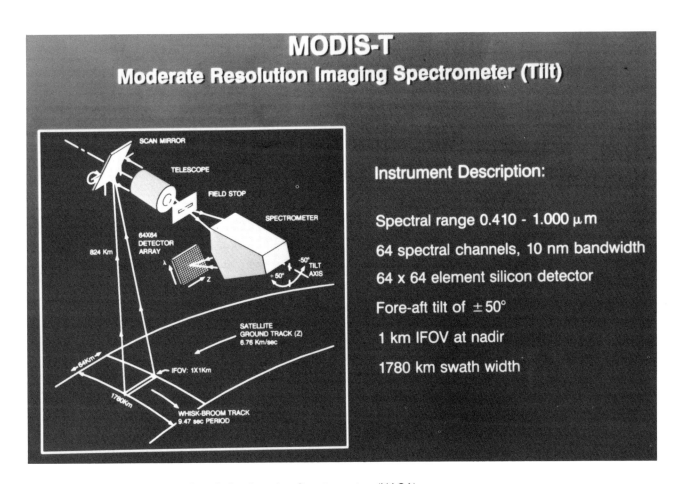

Figure 14. Modis-T Moderate Resolution Imaging Spectrometer. (NASA)

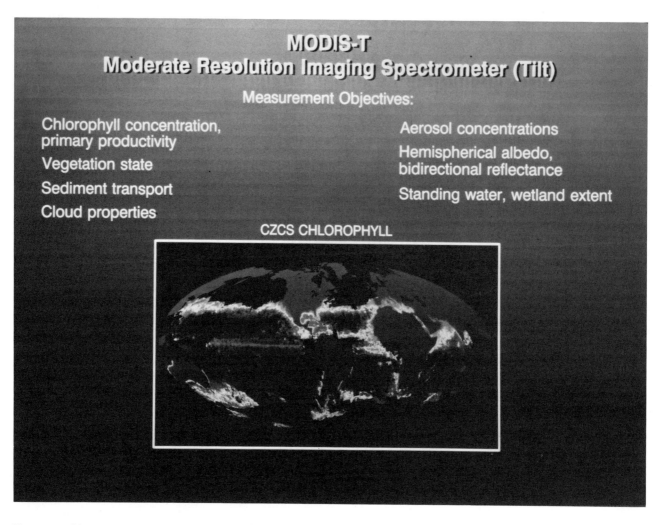

Figure 15. Measurement objectives and type of global map that will be produced by the Modis-T Moderate Resolution Imaging Spectrometer. (NASA)

more data than NASA has accumulated so far to date from all its missions. We will get that every single day. This question of handling of the data may be the greatest single challenge facing EOS. We have two subcontracting firms now examining the question of the data system architecture. We are looking at how we can develop a system to network and make available the data to the scientific community. One thing that is clear is that we must standardize the use of symbols and language so that it can be sent out in a documented and reliable manner. One solution we're looking at is that we employ an artificial intelligence system on board to determine which data is important and which is not.

From the information gathered, EOS is going to be generating maps. We are going to have to learn to put those data bits and pieces together and understand the whole Earth. We will see during this period enormous numbers of global maps using data generated from the EOS instruments. An example would be the Modis-T Moderate Resolution Imaging Spectrometer (Figure 14). Figure 15 shows the measurement objectives and the type of global map that it will produce.

In summary, the concerns motivating the EOS project planning is that the Earth is changing. It is very likely to affect our weather and climate. We have to understand it. We need more data. This will take a large international effort and EOS is very vital and the next logical step.

Business and Life in Space

Joseph Allen, Ph.D.
President, Space Industries International, Inc.
Former NASA Shuttle Astronaut
Houston, Texas

The suggested title — "Business and Life in Space" —seems a little bit tame for my tastes. I would rather entitle my talk, "The Ecstasy and the Agony". The reason for this will become apparent from my remarks.

First the ecstasy — including the joys of space-flight and the ingenious environmental systems we humans have invented to enable such flights. I am going to use some analogies between space travel and the Columbus voyages. There are some interesting similarities — and some very obvious differences — between the voyages of Columbus and the Space Shuttle *Columbia*. One difference is that our spaceships always start out traveling east, not west. The initial speed of a spaceship is probably even slower than the speed of the Columbus ship — about a mile an hour as it moves very slowly towards its launch pad. Some time later though, the analogy breaks down dramatically. For example, a spaceship is not westward bound, but upward bound into and across unknown oceans (Figure 1). The oceans of space are better known now than they were 30 years ago when the space age began, but there is still much to be learned.

I am going to talk some about the life support systems in this machine we call the Space Shuttle and then later about life support systems in a little cocoon that is far smaller than the shuttle. I like to think of it as a cocoon. The more common term is a space suit. A lot of numbers and design aspects of these machines are rather intriguing. For example, the space ship would tend to heat up during the outbound journey, so we evaporate overboard both ammonia and water during the ascent to orbit

in order to keep the crew compartment at a comfortable room temperature.

Another Columbus comparison that intrigues me is related to the story that Columbus's crew was very fearful of sailing to the edge of the Earth and falling off. Now, I think it must be the case that Columbus himself knew that was not going to happen, and Queen Isabella knew that was not going to happen. The intelligentsia of those times knew the world was round, the only argument was how big was it. Columbus felt it was rather small and he loaded his ships with enough food and water to carry him around the small Earth to India. As it turned out this world is not as small as Columbus estimated and he did not get to India. Luckily though he did not have to rely, as a life or death matter, on his closed environmental systems either. He was able to resupply food and water from a "new" continent that we now know as America. Queen Isabella didn't care whether Columbus lived or died anyway. Odds were that she was going to lose her money, but she had bet on his somehow surviving because, if he did, it solved one of her major political problems. As they say, the rest is history. Governments were clearly a little more cavalier about their explorers in those days than we are in this considerably more timid age.

To repeat, we are told that the Columbus crews of the *Nina, Pinta* and *Santa Maria* were afraid of falling off the Earth. We, the *Columbia* crew of the *Discovery*, sailed for eight minutes and our modern technology enabled us, on purpose, literally to fall off the edge of the Earth. This is an absolutely correct statement in physics. When the engines shut down, we are falling and we continue to fall, fall, fall around in orbit. We remain in this perpetual free fall until it is time to come home. I will give you some more intuitive ways to think of free fall in just a moment.

Life aboard a spaceship: we are basically in a rather small cabin. Physically, were we to be in such a small room or mid-deck volume here on the ground, you would find it very crowded indeed, particularly if there are as many as eight people inside. In space, a small volume is not nearly so confining — the reason being of course that people float all over the place. In fact, you can sleep on the walls, you can hover on the ceiling, you can disappear into little nooks and crannies in any direction. Even a small volume becomes rather spacious in space because you are living in all three dimensions. On Earth we humans are confined largely to the area of the floor. This is no doubt why the size of a home or apartment is cited in area (square footage) rather than in volume.

Let me talk about aspects of space food. We eat largely freeze-dried reconstituted TV-like dinners. This, by the way, is an example of the "stow and throw" philosophy that we heard about yesterday. There is nothing "closed-system" about this. When we finish consuming a pre-packed meal, we throw the packaging and the remnant food away. One of the constraining aspects of the "consumables" in the Space Shuttle is that we run out of trash stowage volume fairly quickly. Although we run out of liquid oxygen and liquid hydrogen that give us the necessary electrical energy in a Space Shuttle in a little over a week, we would have to come home in about two weeks anyway because we would literally run out of places to stow the trash. We do not throw trash overboard. The Russians do throw it overboard, however.

Although the orbiter looks like an airplane, it is not like an airplane at all and living aboard an orbiter in space is very much like living aboard a ship. There is no engine sound. By no sound, I mean no constant engine noise, thus the cabin sounds like a modern computer-filled office. The pressure of the crew compartment is kept at one atmosphere, i.e., 14.7 p.s.i. When you go into space, your ears don't even pop. The humidity is also very carefully controlled. Given the fact that we are all from Houston, it is controlled at 100% — well, perhaps at 50%. Life in space is somewhat like being aboard a submarine but with one remarkable difference. There are 10 windows in an orbiter and in many ways, they make all the difference. Even though they are somewhat recessed and thus not easy to look through, very shortly after you arrive in orbit each window is covered with "nose smudges".

Figure 1. The Space Shuttle Columbia on launch from Kennedy Space Center, Florida.

Looking at Earth photos taken through these "windows on the world" the atmosphere appears as thin as an onion skin. It is not hard to believe that one could punch an ozone hole in something only that thin. You of course don't see ozone damage, but you do see the delicateness of the atmosphere with your eyes all of the time. In this photo (Figure 2) we see in one frame the Pinacate Mountains, the area of Biosphere 2, and further up the California coast we can see where Ames Research Center has moved 16 inches closer to San Francisco during the recent earthquake. That is pretty much the scope of what your eyes see from orbit. But the photos you see are only still photos — the scene from orbit is always moving. In orbit we would be traveling at about 5 miles a second, so during the time that we have looked at this photo we would be speeding past the Gulf of Mexico,

Figure 2. Baja California and the west coast of the United States as seen from the Space Shuttle. (NASA)

were we in space right now. I will confess that it is very difficult to do the Government's work when the Government gives you all of these windows to look out of.

Back to more technical things. The orbiter is literally an envelope of Earth with windows. The consumables within the envelope that we run out of first is electrical power generated by hydrogen and oxygen combined in fuel cells. The next thing we would run out of is volume for storing waste. Then we would run out of food. We would also run out of lithium hydroxide containers that are scrubbing the CO_2 out of the atmosphere and ultimately we would exhaust the oxygen supply from which we are breathing but oxygen is used for other things as well. For example, oxygen is combined with hydrogen in fuel cells to generate electricity. A by-product is water that we drink. The O_2 and H_2 when combined in fuel cells also generate heat and the heat is dumped (dispelled) through radiators. On balance we wind up with too much water and we actually must dump water overboard from time to time. When we dump water there is always a fight to get to the window nearest the dump port because the sight is like orbiting the Earth inside a blizzard. The water comes out, immediately freezes and sublimes away, but you are enveloped in a snow storm for just a moment when that happens.

This is the cocoon I spoke of earlier (Figure 3), a person in a space suit. This person is kept warm, supplied with oxygen at 3 p.s.i., given pressure, which is also important so that she can fill her lungs and absorb the O_2 into the blood stream, and supplied with a radio so that she can talk to friends and neighbors. One is typically asked, "aren't you lonesome out there in your spacesuit?" The answer is "no", because somebody is always talking to you. In addition, you feel all bundled up, exactly like when your mom put you in your snowsuit many years ago. You do feel very encumbered and, although it is a comforting feeling, it can be a frustrating feeling at the same time. For example, the minute you put on your helmet you can no longer scratch your nose, or any other part of your anatomy, I might say. If you have a tear in your eye, you can not rub your eye nor does the tear roll out of your eye. It stays and you will spend some minutes looking through the tear drop as though you are under water. Interestingly this tiny bit of our human environment would tend to heat up in the space environment without active temperature control. It is kept cool by evaporating water from a metal plate located in the back pack. The first thing that this suit will run out of is cooling water. We watch very closely the cooling water level and the

Figure 3. Astronaut in EVA space suit. (NASA)

Biological Life Support Systems

minute it gets down to the last bit of water the next procedure is to return to the orbiter. The spacesuit is outfitted with food inside, candy bars that fit down in front of the suit and that you can get to with your mouth. The technique is pull up the candy bar and then bite it off. Don't bite it off and then raise your head. Note that it is very important to remember the correct sequence. It takes astronauts a long time to learn but once you get hungry enough, you have learned; otherwise, you might stay in the suit for a full eight hours with no nourishment. There is also fluid stored inside the space suit and a straw that you can get to with your mouth — the drink could in theory be the liquid of your choice, but the Government does limit the selection to water or Gatorade. They totally ignored my request. As far as other body functions are concerned, once again you must use the only technique available to you when bundled up by your mother in the snowsuit. If you have to go, you just hope the diaper doesn't leak.

There are three satellites in Figure 4 — part of the orbiter, which is a satellite, the communication satellite named *Westar* and a "satellite" named *Dale Gardner.*

Figure 5 is in a sense a demonstration of the zero-gravity which results from the falling around planet Earth. The object shown is not a child's balloon, rather it is a photograph of floating liquid taken inside the spaceship. It is a cola soda. It's a well-known substance, but you have never seen it in this state. When you drink a carbonated soft drink in zero-gravity you notice the bubbles don't know which way "up" is. In other words, the carbon dioxide bubbles don't rise to the surface and pop out because without gravity there is no buoyancy to move the light gas to the surface of a heavy liquid. Once again, there is no up or down here. To dispatch the liberated fluid, you can drink it with a straw or you can just attack it with a wide open mouth.

During the reentry of the orbiter into the Earth's atmosphere — although in this photo it is nighttime outside — we see light from the ion glow caused by hitting the air molecules at around MACH 20 [20 times the speed of sound]. The orbiter comes home

Figure 4. Astronaut Gale Gardner in an EVA from the Space Shuttle with a communications satellite.(NASA)

making enormous "S turns", which is a subject of another lecture. Figure 6 is a photo that I rarely show of the view looking out the back of the ship. The tail would be here, if we could see it. It is the image of the ion glow spilling around behind the orbiter. The glow waxes and wanes, moves and flickers in size, color and intensity. I wanted to show this slide yesterday, Halloween, because it looks for all the world like the most eerie figure. When you first see it, you hope it is the Angel of Good Technology, not the Angel of Bad Engineering. My final photo (Figure 7) taken at the end of the first flight of Space Shuttle *Columbia*, comes with a newspaper headline, "Today a spaceship landed on planet Earth."

I want to go from the ecstasy to the agony of space exploration. The ecstasy, of course, comes from our past space accomplishments; the agony comes from the bureaucratic snarl that is increasingly smothering our potential for future accomplishment. In short, although through our technology we are now in a position to undertake truly astonishing projects, the way our nation's laws are being applied make these undertakings nearly impossible. Indeed, I contend that the greatest

challenge to us space workers is not unraveling and applying the laws of nature to space exploration, but rather, finessing around the ponderous laws that have been put in place by the military, industrial, and bureaucratic political complex. Those are very long words — I don't mean to appear pessimistic about it — but I frankly am very worried about our ability to move forward as a nation and undertake, successfully, many of the projects that we have the technology, the energy and the conviction to do.

I am going to give two examples that are indications of this unfortunate trend. The first example is the current Space Station, and the second is an example that I have imagined just for this occasion. If we look into the night sky we can find the lights of an orbiting Space Station — the *Mir* of the Soviet Union. I would have loved to have brought a picture of the American Space Station. Unfortunately, it's in the form of boxes and boxes of plans that would fill this room many times over. We have no actual Space Station, but the other Space Station is working right now. People are in the Space Station. They are called "cosmonauts." We at Space Industries have an experiment going aboard this Rus-

Figure 5. Cola soda in the microgravity environment of the Space Shuttle. (NASA)

Figure 6. The ion glow spilling around behind the Space Shuttle orbiter. (NASA)

sian station in six weeks. It was the only place we could take it. The Russians have made it very convenient for us to fly with them and we are going to do it.

However, in this country we have a commitment to a Space Station and, indeed, to implement such a project should be nowhere near as difficult as an Apollo project. It is, in fact, not as difficult as the Biosphere 2 because we already have built a practice Space Station. We called it Skylab. It was done with monies left over from Apollo — about a billion dollars. We talked about Skylab in '68 and '69, we constructed it in the early '70s and we flew it in 1973. So a Space Station is not something new. The official International Space Station was committed to by President Reagan in 1984. It was to be flying in 1992 in time for and in celebration of the 500th anniversary of the discovery of America. Thus, when we committed to it, the Space Station was eight years ahead of us. Well, some years

have passed. We are now in 1989, nearly 1990. Where is the Space Station? The best estimate is that it will be ready in 1999 — now 10 years ahead of us. Thus between 1984 and 1989, we have spent over $2 billion on the Space Station and we have lost two years. Because of bureaucratic inefficiencies the faster we go, the behinder we get.

I will add another thought. There are some numbers around which reflect the costs of the early space machines, both what they cost and how the resources were put into them. For example, when you look at the Apollo spaceships, you will see that about forty percent of the cost of Apollo went in either to the hardware or to the wages of the people who were building the hardware. If you look currently at the monies that flow into the Space Station you will see two billion dollars are going into this Space Station just this year [1989]. Two billion. By the end of this year you would think that we would have a piece of metal some place to show for the

Figure 7. Space Shuttle *Columbia* preparing for landing after its first spaceflight. (NASA)

money. We won't. Maybe $500 million will go towards actual hardware, however, of the $500 million, the Congress actually cut $250 million and so that cut comes out of the actual production of the machine. And the remaining 75% of the $2 billion, purely overhead, continues to consume all the rest of that money. The trend is that pretty soon we will be spending infinite money and getting zero product to show for it.

Let me end these thoughts by projecting a scenario which I fear could actually happen. First of all, I join all of my associates from NASA in congratulating and encouraging those of you who are associated with Biosphere 2. This Biosphere 2 project literally feels like NASA felt to us NASA-ites in the old days. Keep after it. It is a wonderful feeling. It doesn't mean you are going to do everything right, but you are at least doing things. Let's assume that there are elements of what you do that, in fact, actually work. We also know that on the 20th anniversary of the Moon Landing our President announced that, in addition to the Space Station, we are going to return to the Moon, this time to stay, and then travel further on to Mars. Wonderful words. I would assume that in response to this challenge our government officials will need information about biospheres.

Let us examine how our government will procure information about biospheres. Government officials will need to lay out a set of requirements. Many workshops, studies and hundreds if not thousands of consultants will be involved. It will take at least a year to lay out requirements for information on biospheres needed in order to undertake a lunar colony. From those requirements, still more committees will derive a set of specifications that contractors will have to meet through contractor developed designs of possible biosphere configurations. These specifications will be formally detailed in a "Request for Proposal for Phase B". The U.S. aerospace industry will respond to this formal request with equally formal proposals — each proposal from a team composed of various large companies and each describing what that team could do in terms of meeting the specifications which in turn will satisfy the requirements.

Let's assume that Space Biosphere Ventures would like to participate in this Government competition. You would have to spend at least $200 K just to submit your proposals in competition for the Phase B study money. This effort would take a half a year of your time. It pains me to predict that your team would also be found "incapable" of studying a biosphere because of many things that the government requires in order for a particular group to be a legitimate government contractor. In any case, the government considerations on who is the proper vendor for a Biosphere G (the Government Biosphere) would unfold for months and months. Years would pass and ultimately a phase C and D contract — to build Biosphere G — would be given to the winners of the contract. They would be Lockheed, Boeing or McDonnell Douglas — well-known aerospace companies, but extraordinarily expensive. I am certain that by the time of those awards more tax money would have been spent on studying Biosphere G than you are going to spend here on doing Biosphere 2. And years will have passed. By the time Biosphere G is officially undertaken I hope to goodness that you will be well along on your Biosphere 8 and, furthermore, that Biosphere 8 will be either outbound or already located somewhere in the far reaches of our amazing Solar System.

It is very distressing and I realize I don't sound optimistic. But in the long run I am very optimistic, partly because of places like Space Biosphere Ventures. If our government is unable to make progress, that does not mean humanity is going to be unable to do it.

A Personal History of the Human Exploration Initiative with Commentary on the Pivotal Role for Life Support Research

Wendell Mendell, Ph.D.
Chief Scientist, Lunar Base Studies
Johnson Space Center

In 1961, President Kennedy announced an initiative to land a man (you would now say a human) on the Moon and return him safely to Earth. It was a political statement, and everybody knows that. However, it represented a straightforward, simple mandate to an engineering organization such as NASA. Engineers really like, culturally, to receive sets of requirements and turn out a product designed to satisfy those requirements. Apollo, in some sense, was an ideal example. The task to accomplish was specific: land a human being on the Moon, and get him or her back safely back to Earth. Within that context, they could build transportation systems and develop the technology necessary to accomplish that.

It was a very happy time. NASA had a lot of money and a lot of support from the nation. The task was a very important thing to do, so it was pleasant duty and a lot of fun in a new organization. As the Apollo program actually started to come to fruition in the late 1960's, questions arose as to what to do next for an encore and how to continue this line of discovery. There were things called Apollo applications, aimed at extending stay times on the Moon and so on.

In fact, President Nixon in 1968 asked Vice-President Agnew to head up a Space Task Group to draw up a set of plans beyond Apollo. That report is really interesting to read because it talks about space stations, lunar bases, and bases on Mars. If you look at the time-lines, funding curves and schedules in that report, we should have been on Mars about 5 to 8 years ago; by now we should have a permanent base up there, along with the permanent base on the Moon.

Well, that didn't happen. What really happened, of course, was that when this plan was taken to the Office of Management and Budget by NASA's Administrator, Tom Paine, he was basically told, "No way — we've got budget cuts to deal with."

I have read the history books; I wasn't there. The history books and the analyses say that Dr. Paine just was not aware of the level of political difficulty he was in at the time he presented his plan. He went directly to the President to argue that this ought to be done, ought to be accomplished, and he lost. And when you go directly to the President and lose, you can really lose big.

President Nixon had other priorities on his mind — the Viet Nam war for one, and the Great Society program that Lyndon Johnson had put through the Congress which now had to be funded. Those two things were getting a lot of attention in the Congress. Funding for the space program seemed like a luxury. After all, we did beat the Russians to the Moon; perhaps it was time to move on.

NASA backed off its plans for bases on the Moon and on Mars and, instead, proposed an Earth-orbiting manned Space Station. That proposal was cut back in negotiations with the Nixon Administration until the only piece left was an Earth-to-orbit logistics vehicle, which came to be called the Space Shuttle.

The design philosophy of the Space Shuttle was to lower launch costs to orbit by analogy to the operation of airlines: build a small number of vehicles and keep them flying through rapid turnaround on the ground. According to testimony to Congress in the early 1970's, the Shuttle would fly 50 times a year and, therefore, the launch cost to Earth orbit would fall dramatically. It didn't work that way, and the reasons were not necessarily all technical. Some design decisions on the Shuttle program reflect requirements set by the military (whose support was needed in Congress) and by restrictive budgets.

If you look at the history of NASA during the 1970's, you find it preoccupied with making the Shuttle work and with planning a Space Station. The whole context of space exploration set by the plans of the Space Task Group somehow got dropped from the corporate memory. Gradually, activities called "Advanced Planning" in NASA were eliminated.

In 1980, the Space Shuttle, behind schedule and having difficulties, still had not flown its first mission. The new Reagan Administration had come in under a banner of austerity and less government, intending to cut spending and eliminate programs. Of course that sent a tremor of fear through the Federal bureaucracy, of which I am a part.

The new NASA Deputy Administrator, appointed in the very beginning of the Reagan Administration, also had very definite ideas about priorities within the space program. He made known his opinion that the limited resources of the Agency ought to be put into making the Shuttle work. That might mean cutting back on other things, among them the planetary exploration program. In fact, the rumor mill said that the planetary science budget was going to zero over three years, leaving the Agency to concentrate on manned exploration. At some point in time, once the Shuttle was an operational vehicle, we would resurrect the planetary programs, like Lazarus, and start exploring the solar system again. That scenario sent shock waves through the scientific community (of which I was a part), and a number of things happened.

Our particular group, being civil servants, couldn't participate in activities like Political Action Committees being formed by scientific societies. We looked at strategic planning taking place in the NASA Headquarters Planetary Program Office in

response to this and felt it was flawed. We were forced to reexamine our performance as a scientific research group to understand how to restructure and set priorities to remain competitive in what would be a highly restricted funding environment. We formed some internal committees to examine various options.

One of the things we revisited was the Lunar Polar Orbiter, a mission we had proposed in the early 1970's but which had never flown. In its original incarnation it was to have been a very minimal mission, intended to be launched on a Delta rocket to the Moon. However, under the guidelines of the early 1980's such a mission would fly on the Space Shuttle; the increased payload capacity of the Shuttle over a Delta would give the whole mission a great deal of capability. Thus, it seemed like a good idea to look into the technical issues associated with launching a lunar satellite from the Space Shuttle and become acquainted with a very broad spectrum of opportunities and possibilities in this "future space program." We could accomplish this at the Johnson Space Center by walking across the campus and talking to the engineers.

Those of you who don't work in NASA may not be aware how little working level communication there is between the engineering side and the science side. Even though located at the Field Center for manned flight, as a scientific organization, we rarely dealt with the engineers in the manned programs. Our time was spent with individual research projects and conferences about the planets and so on. However, at this point we had a real need to talk to those guys.

We made a couple of rather startling finds (to us) about NASA's future. We, of course, knew about the Space Shuttle, but we learned that it was called the National Space Transportation System, a name which implied there was something more to it than just the Shuttle itself. A Space Station was on the drawing boards and, back in the dusty corners, engineers were talking about spaceships called Orbital Transfer Vehicles (OTV) that could carry payloads from a Space Station into high orbits. These hypothetical vehicles (which would exist in the mid 1990's) would be able to take payloads not only to places like geostationary orbit but also literally to lunar orbit, without any changes in their propulsive capability.

That connection lit a great light bulb in my brain — for scientists who were interested in the Moon, the late 90's were going to be a real Golden Age, a new era of discovery. As this transportation capability came into service, our little dinky payloads could go up there all the time, collecting data on the Moon. Probably there would be a base on the Moon and other activities beyond low Earth orbit.

Exploring these possibilities within NASA, we found that such ideas were considered crazy stuff. NASA was having plenty of trouble getting a Space Station — if we were to talk about the Moon, then Congress will never allocate money for the Space Station. This type of thinking seemed to imply that the space program would never be more than a short term activity. We could not accept that, so we started an effort to map the long range structure of the space program.

At the very beginning of this activity, in early 1982, we began a collaboration with the Los Alamos National Laboratories, where we discovered (accidentally) that similar discussions were taking place. Jointly our group and the Los Alamos group developed a set of premises about space development, and then began expanding the dialogue to senior people in the space business who also had concerns about the future. Principally, we concluded that it was inappropriate to think of a Space Station or a lunar base as simply "a next logical step" as NASA moved from one project to another. What we were really looking at was the very beginnings of a process of moving human beings off the planet into the solar system.

Unfortunately, characterizing NASA programs in terms of grand, historical processes was not acceptable to the internal bureaucracy. That kind of talk was considered fantasy, given perennially tight budgets. Thus, our first step was to get the idea across that it was actually okay to talk in such terms. In other words, we needed to legitimize the concept of human exploration of space in general and the idea of a lunar base in particular. To that end we employed a number of tools such as workshops, lectures, sessions at technical conferences,

and articles in the public media. I have a number of clippings from 1983 and 1984 that talk about lunar bases as if they were part of the NASA pantheon even though the agency itself was doing almost nothing in the field.

This orchestrated legitimization process led to a conference in 1984 that we held at the National Academy of Sciences resulting in the book that some of you have seen, entitled *Lunar Bases and Space Activities of the 21st Century.*

In late 1984 and in 1985, other people began arguing for piloted missions to Mars. Carl Sagan and Jack Schmitt were both lobbying James Beggs, the NASA Administrator, about missions to Mars for totally different reasons. Sagan wanted to go to Mars with the Russians as a world project to resolve global political tensions through technological cooperation. Jack Schmitt thought the Russians were going to do it and leave us behind in the dust.

Beggs realized that NASA had not thought very much about going to Mars in the Shuttle Era. In late 1984 he called the Director of the Johnson Space Center because he knew that a group there had been working on some of these ideas. Beggs asked that they put together a NASA-wide study to review how we would reply to the President if he said, "Go to Mars!" That study was done in about six months in the first half of 1985 and was published in 1986 as a Marshall Space Flight Center document.

At about that time, Congress mandated the President to appoint a National Commission on Space (NCOS) to look at the long range future. President Reagan finally did that, after some delay, in early 1985.

These activities began to gather some momentum and start to be recognized in some sense, but future goals never got on the charts (NASA viewgraphs). For example, in the justifications for a Space Station, the top reason was microgravity research or commercial development or better ball bearings. An engineer once told me after one of my presentations that I was talking about transportation of human beings into space and that was the seventh of seven reasons for the Space Station. I acknowledged the problem but emphasized that I saw the Space Station not so much as a research laboratory but rather as a stepping stone toward permanent human presence in space.

By the end of 1985 a new paradigm for the Space Station was beginning to emerge in a lot of people's minds, but it was still not manifested in any way in NASA officialdom. That was the state of affairs in early 1986 when the Challenger disaster occurred. The NCOS had its report ready to present to the President, and this terrible thing happened.

However, one of the results of the Challenger explosion was the bursting of the bubble of the Shuttle fantasy in the High Councils. At the time the system was working toward 16 flights a year and not 50 any more, but even that goal was beginning to be seen as a tough problem. A realization was dawning that the vehicle was essentially a research and development tool, not an operational system. There was real risk involved. NASA suffered a great deal of examination and critique.

Nevertheless, the Challenger tragedy generated a tremendous outpouring of public support. I think people at the top of NASA were surprised by the positive feelings because they live in a highly political environment in Washington, DC where they are constantly beset by negative and critical views. That is really all they hear. Do you want homes in space, or do you want homes for the homeless? This is the trade you are making. I think that even though they may have known intellectually that there were people who loved the space program, the support that came out in the national media and everywhere was something of a surprise. And it encouraged them to think more about the future.

The NASA Administrator accepted the NCOS report and later that year asked Sally Ride to study possible future initiatives for the U.S. space program. Neither the NCOS operation nor the Ride Study was large enough to have an independent technical staff. Both groups had to draw on preexisting information, and almost all of the recent stuff was our bootlegged work of the previous four or five years on lunar bases and the Mars mission study. Most of it was very conceptual, but it became the basis for much thinking.

The National Commission on Space report, *Pioneering the Space Frontier*, had a very grand vision that philosophically broached the question of human settlement of the solar system. Later, that idea appeared in the Reagan space policy of February 1988. Thus the report was very important, even though people thought of it as a "blue-sky" study destined to molder on the shelf. It created an important philosophical basis for things that would come later.

The Sally Ride study posed four grand options for NASA: Mission to Planet Earth (which you heard about last night); Robotic Exploration of the Solar System; Outpost on the Moon; and Piloted Missions to Mars. Those sound like four distinct choices but, in reality, they are different scales of activity. As I said earlier, I have to agree with Mel Averner that Mission to Planet Earth is really not in the same league as Outpost on the Moon or Missions to Mars. When you really look at a Mission to Planet Earth program, even on as grand a scale as has now developed, it is something like a factor of five or so smaller than a lunar base program. A Mars program started from scratch is probably another factor of two larger than a lunar base program in terms of expense.

NASA formed an organization called the Office of Exploration (OExp) to continue the work of the Ride Report and to explore these options in more detail. However, that organization was chartered to study mainly the Moon and Mars missions. It was assumed that the planetary exploration element is really being taken care of very well within the current offices of NASA. The Mission to Planet Earth is really something a little broader and larger than NASA, not necessarily a NASA program and not of the ultimate scale of human exploration of the solar system.

Office of Exploration began its work, I guess, in late 1986. In December of that year, the NASA Administrator circulated a memo to all NASA employees declaring that one of NASA's major goals was to expand human presence beyond the Earth into the solar system. Little notice was taken of that statement, but it was echoed about a year later in the February 1988 space policy issued by the Reagan Administration. That was really important for it allowed NASA to actually think about human space exploration in terms of long range goals. You have no idea how important it is to a bureaucratic government organization to be given permission to think about strategic issues.

Given that permission, the Office of Exploration took it upon themselves to come up with a long term strategy for the space program. However, they first felt a need to educate themselves about the implications of various choices. Let's take "Outpost on the Moon", for example. A lunar base could be a Chevette or a Cadillac. You can just put people up there, plant the flag, and bring them home; or you can establish the beginnings of communities.

What about bases on Mars? Do we go there, land a couple of places, and say, well, we did that, i.e., "Little Jack Horner sat in a corner, stuck in his thumb, pulled out a plum, said, 'What a good boy am I'?"

There are all sorts of scales to these things, and we don't always understand what it means to adopt one or the other of these scenarios. The Office of Exploration wanted to provide recommendations, alternatives in the early 1990's. The target date for a final recommendation was 1992: the 500th anniversary of the discovery of America, the International Space Year.

Rather than sit down and try to develop a plan immediately, they chose to do a series of homework problems. The approach was to formulate a series of problem statements of the sort you might find at the end of a textbook chapter. Solving these exercises would give insights to the workings of the methodology and to the implications of various decisions. They were very careful to refer to their work as "case studies", not scenarios. The word "scenario", as Gerry Soffen pointed out earlier, implies that you have converged to a plan. If the press thinks you are developing a scenario, they assume the it is the first draft of the final plan. In reality, we were doing practice runs, and they were called case studies to emphasize that.

In that process they arrived at some generalizations from these case studies. One was a classification called *Human Expeditions,* or "flags and footprints" as it is called informally. A human expedition means that you are just demonstrating capability and, perhaps, collecting information. The Lewis and Clark Expedition explored and reported back but didn't leave behind any outposts or settlements. If a facility or some scientific experiments is established which can be revisited, we refer to it as an outpost. An outpost does not have to be permanently staffed.

Finally, there is a rather revolutionary notion called *Evolutionary Expansion* in which a permanent presence is established with an intent to grow to self-sufficiency in an economic or material sense. This latter concept begins to transcend a simple programmatic decision and has the potential to inaugurate a historical process.

I came to the conclusion some time ago that the inevitable maturation of space transportation technology implied that the human race was ready to begin permanent settlement off the Earth. The only question in my mind was whether Americans would be leaders in this process, whether our values and ideals would become part of the foundations of space-faring societies.

A vague, philosophical idea like evolutionary expansion is difficult to deal with in an engineering-oriented organization such as NASA because it doesn't immediately lead to a set of requirements against which engineers can design machines, or give you a series of steps toward a specific goal. So I sat in interesting meetings, watching the mind trained in the engineering culture struggle with really philosophical issues where you had to derive what you wanted to do from a general cultural imperative. That was a very difficult exercise within this organization, but some good progress was made, ultimately.

One case study, or problem statement, investigated was a human expedition to Phobos. By landing on Phobos, you don't have carry the mass with you to land on Mars. The objective is to get somebody into the Martian system as quickly as possible with technology that you have at hand. That was the intent of studying that issue. It didn't necessarily mean that they were trying to advocate landing only on Phobos. In fact, the case study included robotic exploration of the surface of Mars, using teleoperation from Phobos as a base. This is an old idea that Fred Singer came up with as part of his PhD thesis, I think in the 1950's. It may even predate Fred, I don't really know, but he certainly popularized the idea.

Doing this problem forced you into some on-orbit operations, but required only modest mass in low Earth orbit (LEO) and less time for program development — characteristics which made it an important case to understand. Of course, a lot of people thought it was just crazy to go to the Mars system and not land on the planet. Therefore, you also had to include the case study involving human exploration of Mars.

In that study it became clear that the Space Station was needed for assembly in LEO because you can't bring up everything at once in big pieces. On Mars we have robotic exploration of the Martian moons instead of vice versa. Much technology development and operations experience was needed at the Space Station, particularly research in life sciences.

As I sat in the meetings in 1985 for the Manned Mars Mission study, I had realized for the first time how much the decisions related to the Mars missions were driven by our ignorance of the life sciences. Our limitation was not engineering or ability to design the transportation systems — our limitation was our understanding of the human being and how that human being might adapt and perform on a three year round trip. That kind of experience is like the old sailing voyages around the world in the 16th century.

Mars landing requires a lot of vehicular and space systems infrastructure within a launch window that opens only once every 26 months. A huge spaceship has to be built in LEO, and if you happen to fall behind schedule a few weeks, you maintain it there for another 26 months. An enormous management operation is involved just to meet that schedule. Something like 500 tons of propellants alone have to be shipped to orbit once you have built the spacecraft — an imposing challenge, considering how we do business today.

The case study designed to evaluate lunar activity was taken to be a science outpost on the far side, using an optical interferometer located on the lunar surface — an idea Bernie Burke published first in a book which I edited, *Lunar Bases and Space Activities of the 21st Century*. The lunar surface is extraordinarily stable platform and therefore a unique location for the elements of an optical interferometer. We are examining now the broader and broader categories of scientific experiments that are possible on the Moon but not possible on the Earth or in orbit somewhere.

Of course, the more traditional sort of concept is the Arecibo style radiotelescope in a crater. This idea appears in NASA viewgraphs as early as 1971. I have always thought this scale of project was pretty ambitious and only a nice thing for artists to draw. However, at a symposium on Astronomy from the Moon, held in 1986, Frank Drake pointed out that the Arecibo telescope in Puerto Rico is built suspended from only three pylons on the sides of the crater. All the structure is supported by cables. When you realize that, the civil engineering problem doesn't seem nearly so difficult.

The lunar case study focussed on operating a long duration science facility on the Moon that would be man-tended but not necessarily permanently manned. Clearly, substantial scientific capability could be put on the Moon within a relatively short time. Massive human presence is not required, but human interaction would greatly enhance the performance of the installation in terms of maintenance and change-out of instruments. The mass launched to LEO to do this kind of operation on the Moon is much less than for Mars missions.

Finally, there was a case study called evolutionary expansion. A long time was occupied in even getting a grip on what that meant. It was not studied as deeply because there were so many false starts over its formulation. Nevertheless, one of the ideas very prominent in evolutionary expansion was to somehow use the Moon as an outpost early on to build your infrastructure, test your systems, and learn how to live on planets. It might even be possible to increase your ability to operate in space with oxygen production on the lunar surface. Thus, lunar activities really become a building block on the way from Space Station to Mars to the rest of the solar system. This idea of achieving plateaus or "terraces" in capability and technology has often been advocated by Peter Glaser.

Now, that turns out to be the idea that is manifested in President Bush's speech of July 20, 1989. A lot of people, particularly reporters, complained to me that the speech was "wimpy" because the President didn't give any schedules or details or cost figures. I disagreed and, in fact, thought it to be extraordinarily important because, as Lee Tilton of Stennis Space Center said to me last night, it cut off almost all the branches from this vast decision tree that NASA likes to build. NASA has the idea, and probably rightly so, that it should not make policy decisions. NASA only can provide options to someone else, presumably the President, who will make a decision. All these studies going on inside NASA are suddenly now coming to a screeching halt, and we can really start to focus on specific tasks and accomplishments.

I personally believe that the approach enunciated by the President is the right one. It is one that I have been talking about for a few years, anyway. This way, you end up with a fairly complex infrastructure including lunar surface activities, (maybe manufactured propellants), science laboratories, and vehicles going to Mars. Most importantly, there is an interconnection between things that happen in planetary exploration and things that happen on the Moon. We have sort of a building block approach.

The Evolutionary Expansion case study carried out space development and exploration in a gradual buildup through the Space Station to the surface of the Moon. As I mentioned earlier, my experience with the Manned Mars Mission studies in 1985 persuaded me that the critical path decisions in the Human Exploration Initiative require prominent programs in life science research. The role of the Space Station ought to be to address these issues. The concept of the research laboratory in space in materials science could be satisfied by Joe Allen and his crew with the Industrial Space Facility or its NASA-generated generic equivalent. Astronomers and Earth-observing scientists have

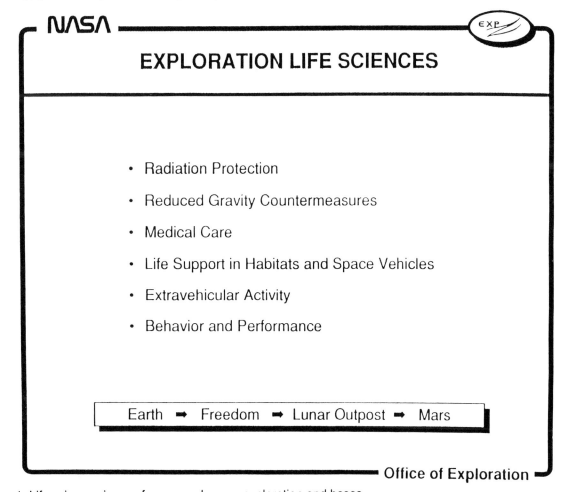

Figure 1. Life sciences issues for manned space exploration and bases.

platforms for their work which are probably more suitable than a vibrating, dirty manned Space Station.

Finally, there is the new concept of establishing permanent infrastructure on a planetary surface. Among the communities that I have been able to get interested in this latter point are the civil engineers and the process industry. As I have pointed out to them, historically, you have explorers who open the frontiers and who are the demigods and celebrities. The builders and settlers come after them. To my mind, there is no fundamental reason why the Space Station has to be built by rocket scientists rather than civil engineers.

A more obvious case is a lunar base where you have construction, manufacturing, processing, and general logistics support taking place. If such a facility were being designed and constructed on the Earth, you would not find NASA involved. For this kind of work you go to Bechtel or Shimizu in Japan or Brown & Root or some other constructor-engineer company. They have the relevant experience but are not now involved in the space program.

When we describe these surface infrastructure elements to those companies, their reaction is that it is a piece of cake but flying to the Moon is impossible. When we go to NASA the reaction is that getting to the Moon is straightforward but building that stuff is impossible. There is no experience in either community that gives confidence in the unfamiliar element. We are trying to close this gap, and it has been closed to some extent within the Office of Exploration.

I have pulled out are a few charts from standard NASA packages that list life science "tall poles" (Figure 1). We can see issues in medical care, zero or low gravity countermeasures, artificial gravity, radiation, life support, and human factors, which is often ignored in NASA. Crew interactions are very much more, I think, an integral part of the Soviet program. They have more concern with these things than NASA does, particularly crew psychological relationships. Extravehicular activity is another question, which is as much a space suit technology issue as it is a human issue.

All of these things begin with the Space Station

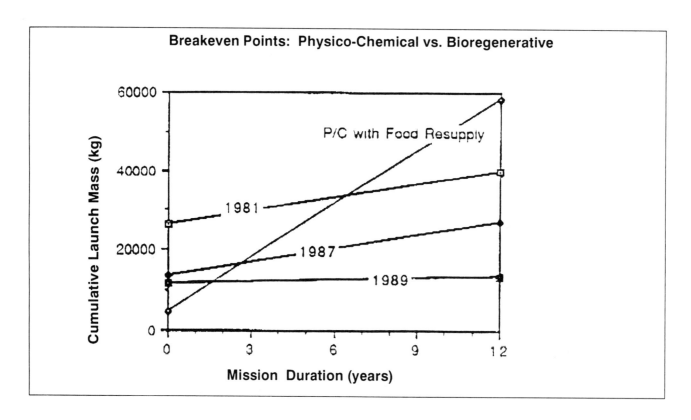

Figure 2. Breakeven points for physico-chemical vs. bioregenerative life support systems for space missions.

Freedom, which gives you the long duration experience, life science research, advanced technology, and so on. Notice how high these Space Station research issues fall on the chart. That is a real change that comes from working on a real problem rather than from generation of rationale for the Space Station by making it up, as in some sense was done in the original proposals.

In the Lunar Evolution concept you want to try to understand the requirements for permanent, self-supporting facilities. Those are really important words: "permanent self-supporting." You also want the capability to be a learning center for long duration planetary missions. So that is a good analogy to what you guys are doing here at the Biosphere 2 Project, developing a learning center and working to be self-supporting and so on.

I have some other charts that were given to me by Barney Roberts, but were given to him by Hatice Cullingford, who has been working on some of the CELSS requirements. These are really things which Mel Averner knows more about than I do. I have run overtime, so I will just pull out a couple of the major ones.

I want to ask you about this chart, Mel, because I wonder about it. This chart (Figure 2) shows that we have learned enough about CELSS during 1981 to 1989 that our crossover point for replacing physical, chemical regenerative systems becomes mission durations on the order of a year. Is that correct?

> *Mel Averner: "First, like you, I had some questions about that. The calculations are based on a study done by Lockheed. I recently talked to a Boeing Aerospace person who independently did the same study from their own point of view and they came out with exactly the same results of a break-even point of about one year. So I now have a good deal more confidence in the report."*

That conclusion is an important one, and new to me. It is important because that implies that projects like lunar bases ought to start investing more in this technology and put more emphasis on it than in the past. It really makes bioregenerative life support a major issue of technology development. I am not sure that this knowledge is reflected in the current planning that is going on. This is sort of new information, so it is something we really need to work on in getting it into the NASA plans.

Another question is whether we can take and derive the lunar habitat from the Space Station *Freedom* with some new technology and then upgrade it to 8 to 12 people, using more and more CELSS technology. I think the pathway that we choose from here through this will be extraordinarily important to efficiently and quickly provide the capability for human beings to live and work in space. Thus, closely connected with all these human exploration goals are some very important and exciting requirements for the life sciences.

It is very sensitive to talk about the report in preparation to the President right now (The 90 Day Report on the Human Exploration Initiative, authored by Aaron Cohen, NASA Johnson Space Center), but the thinking is couched in terms of an emplacement phase, a consolidation phase, and the utilization phase both on the Moon and on Mars. Much of the technology emphasis is on the Moon because we want to develop these systems, make sure they are reliable, and make sure they work in a low gravity environment before we entrust peoples' lives to them all the way to the planet Mars. The Moon is a more forgiving place due to its accessibility by the transportation system.

I would like to conclude with a reflection. This is a magical time, when we have an opportunity within the space program to embark on a truly grand and historical process of human exploration of the solar system. If we can figure how to do that within the constraints imposed by our society and the international environment, there is an opportunity — particularly for the younger people here — to be part of one of the grandest occurrences in the whole history of the human race. We can actually talk about the beginnings of a multi-planet species. The important issues are the ones that Joe Allen raised in his talk — not necessarily the technical ones, but those having to do with the institutional and management structures. Those issues are not as clear and easy to address as are the nuts and bolts, the calculations and the physics.

Speakers Biographical Information

Dr. Gerald A. Soffen was Project Scientist for the Viking missions to Mars, responsible for all of Viking's scientific investigations from orbit, during entry and on the Martian surface. He directed over 70 scientists in the first space mission to successfully perform unmanned experiments on the surface of another planet. Chairman of the Science Steering Group, he was the principal scientific advisor to the Viking Project Manager. He managed biological instrument development at Jet Propulsion Laboratory and served as Chief Environmental Scientist at NASA's Langley Research Center, work which included theoretical modeling, lab experiments, ground measurements and remote satellite sensing. At NASA Headquarters, Gerry was the Director of Life Sciences and helped launch the CELSS project in 1978. His publications deal with biomedical problems in space and the search for life on Mars. Currently, he is Associate Director for Program Planning in the Space and Earth Sciences Directorate, and Senior Project Scientist for NASA's Earth Observing System, an important tool in the Mission to Planet Earth, to increase understanding of the planetary biosphere. He holds a a doctorate in biology from Princeton University.

Mark Nelson is Chairman and CEO of the Institute of Ecotechnics, a London-based ecological development institute which conceived and consults on the total systems ecological management of the Biosphere 2 project. A graduate of Dartmouth College, Mark is a director of the Savannah Systems Pty. Ltd. project in Western Australia which is developing improved pasture and plantation methods for semi-arid tropical regions. Mark also operates experimental arid land rain-catchment orchards in the Southwest of the United States. He is co-author of _Space Biospheres,_ authored a presentation of the Institute of Ecotechnics conceptual model for ecological management in _Man, Earth and the Challenges,_ and was a contributing editor for _The Biosphere Catalogue,_ as well as authoring the chapter on closed ecological life support systems for the International Space University textbook, _Fundamentals of Space Life Sciences_ scheduled for publication this year by MIT Press, and numerous papers relating to the Biosphere 2 project. He organized and chaired the First International Workshop on Closed Ecological Systems sponsored by SBV and conducted at the Royal Society in London in 1987, and was co-chair of the Second International Workshop held in September 1989 in Krasnoyarsk, USSR.

Dr. Thomas Paine served as Administrator for NASA during the first seven Apollo expeditions, including the first lunar landings. In 1985 President Reagan appointed him Chairman of the National Commission on Space, a panel created by Congress to develop civilian space goals for the 21st century. Tom has been awarded numerous international awards, including the John F. Kennedy Astronautics Award of the AAS, the Konstantin Tsiolkovsky Award (USSR), and NASA's Apollo Achievement Award, Distinguished Public Service Medal and Distinguished Service Medal. His extensive administrative experience includes service as Senior Vice President for Science and Technology with General Electric Research Laboratory, and President and Chief Operating Officer with Northrup from 1976-1982.

John P. Allen is Director of Research and Development for the Space Biospheres Ventures Biosphere 2 project, which commenced in 1984. He participated as test subject in the first manned Biosphere 2 Test Module experiment in September, 1988. A graduate of Colorado School of Mines in metallurgical engineering and having received his MBA from Harvard Business School, he served as project engineer on regional development projects in Iran and West Africa with David Lillienthal's Development Resources Corporation. As a metallurgical engineer, he headed a special metals team with Allegheny-Ludlum Steel Corporation. John was a founding member and serves as consultant to the Institute of Ecotechnics, an international ecological development firm which developed the initial conceptual model for Biosphere 2. He was scientific editor of _The Biosphere Catalogue,_ co-authored _Space Biospheres,_ as well as authoring numerous papers and presentations on the Biosphere 2 project for scientific conferences.

Abigail Alling is a marine biologist responsible for coordinating the design, species selection and collections for the marine and marsh biomes of Biosphere 2 with Dr. Walter Adey of the Marine Systems Laboratory of the Smithsonian. A graduate of Middlebury College, with a Master of Science degree from the Yale School of Forestry and Environmental Studies, Abigail has received the International Cetacean Society award for her studies of marine mammals, especially dolphins and blue whales — including a World Wildlife Fund project with Dr. Roger Payne setting up a marine mammal protection program in Sri Lanka. She has directed the research program of the *R/V Heraclitus* of the Institute of Ecotechnics since 1986, including a project to release two captive dolphins successfully back to the wild, and a two-year expedition to circumnavigate South America and voyage to the Antarctica Peninsula — an expedition which focussed on studies of whales, dolphins, and marine ecosystems. She also participated as test subject in the Biosphere 2 Test Module for five days in March 1989 during the second Human in Closed Ecosystem experiment.

Carl Hodges founded the Solar Energy Laboratory at the University of Arizona to create new types of low-temperature, multiple effect solar distillation technology applicable to desert agriculture. In 1967, that venture developed into the Environmental Research Laboratory (ERL) of the University of Arizona. ERL has been responsible for the research, development and implementation of many innovative environmental and agricultural technologies, including seawater irrigation, controlled environmental agriculture and aquaculture projects in the United States, Mexico and Middle East. ERL designed the Land Pavilion of the EPCOT Center of Walt Disney World which routes visitors along a spectrum of global agricultural regions. Through its commercial affiliate, Planetary Design Corporation, ERL has been contracted as consultant to the Biosphere 2 project for some aspects of scientific engineering and as biome design consultant for the intensive agriculture biome.

Dr. Roy Walford is one of the leading researchers in the United States in the fields of gerontology, the immunological theory of aging and lifespan extension. A graduate of CIT and the University of Chicago Medical School, Roy has been a Professor of Pathology at UCLA Medical School since 1966. He has authored over 250 scientific articles, receiving numerous awards for his work. His six books include *Maximum Life Span, The Immunologic Theory of Aging,* and *The Isoantigenic Systems of Human Leukocytes, Medical and Biological Significance.* Roy is a member of the National Academy of Sciences Committee on Aging and was a delegate to the White House Conference on Aging. He has served on the Biosphere 2 Review Committee since 1985, and is the Biosphere 2 Biomedical Program consultant, as well as Chief of Medical Operations for SBV. He has published several papers on the biomedical program for Biosphere 2. Roy is also a candidate for the Biosphere 2 research team to reside in Biosphere 2 for a two year period beginning in the fall of 1990.

Dr. Maurice Averner is Program Manager of the NASA CELSS (Controlled Environmental Life Support Systems) and Biospherics Programs at NASA Headquarters in Washington D.C. CELSS is directed towards the economical production of key life-support functions with three outlined phases: plant growth, food processing and waste processing. The NASA Biospherics Research Program is utilizing existing data banks, remote sensing techniques and field research to create mathematical models that can predict biospheric behavior under a given set of perturbations. With an interdisciplinary background including degrees in biology, public health and law, Mel has been Associate Professor of Systems Research at the Complex Systems Study Center of the University of New Hampshire. Mel served as research associate at NASA's Ames Research Center before coming to NASA Headquarters. Mel's publications include theory and practice of closed ecological systems, and computer modeling of such systems.

Dr. Bill Knott directs the NASA Breadboard Project at Kennedy Space Center, where a Project Mercury pressure chamber has been converted into a CELSS plant growth chamber where air and water are recycled and prospective space crops are tested. With a background in both environmental science and botany, Bill's work includes both micro- and macro-systems. Plant physiologist and biological sciences officer at Kennedy, Bill is responsible for the environmental monitoring program of the 200,000 acre Cape Canaveral nature reserve, including some of the best protected mangrove ecosystems of the Atlantic coastline. Bill serves as manager of the Kennedy Space Center Life Sciences Support Facility and is responsible for coordinating all biological ground-based and spaceflight research. He is principal investigator for all its bioregenerative life support systems research.

Dr. David Bubenheim is a Research Scientist in the CELSS program at the NASA Ames Research Center at Moffet Field, California, directing a development program to include higher plants in an integrated CELSS system. He has worked with CELSS research since 1983. A graduate of Delaware Valley College and Virginia Polytechnic Institute, he did his Ph.D. studies working with Dr. Frank Salisbury at Utah State University on crop physiology in CELSS. He did post doctorate work at Purdue University on experimental investigations of such systems.

Dr. Bill C. Wolverton has, for the past 20 years, directed pioneering research into the use of vascular plants and microorganisms for treating domestic and chemical wastewater, converting sewage into potable water, removing toxic chemicals from drinking water and purifying air from industrial exhausts and inside energy-efficient offices and housing. Recently retired from NASA's Stennis Space Center, Bill consults internationally on applications of his innovative technologies to residential communities of up to 75,000 people, paper mills, chemical producers, aircraft factories and others. Bill consulted with Space Biospheres Ventures' researchers on the application of his wastewater treatment systems for the Biosphere 2 Test Module and in design of some of the air and water purification systems for Biosphere 2. Holder of numerous patents, he has been honored by several awards including the Federal Environmental Engineer of the Year Award in 1983. In 1988 he was one of the first five elected to the U.S. Space Foundation's Space Technology Hall of Fame. Bill holds degrees in chemistry, microbiology and environmental engineering.

Anne H. Johnson is Manager of Research Microbiology at the NASA Stennis Space Center. Anne has worked with Bill Wolverton as a research microbiologist in the development of innovative recycling technologies and currently directs the on-going studies at the Stennis Space Center. She holds bachelors and masters degrees in microbiology and is completing work on her Ph.D. at Louisiana State University. Her research interests include microbial ecology, plant-microbe interactions and the metabolism of sulfur- oxidizing bacteria. Among the test-beds which Anne manages at Stennis are sealed plexiglass chambers for the testing of removal of trace gas contaminants by plant/filter mechanisms, and a prototype "BioHome" which integrates many of these innovative technologies into a comprehensive system.

Dr. Joseph P. Allen is President of Space Industries, Inc., with responsibility for the marketing of the Industrial Space Facility and other company services. Previously, Joe had served for 18 years as a NASA astronaut. He flew on two Space Shuttle missions — on STS 51-A successfully completing the first space salvage of two communications satellites, and on STS-5 which deployed the first communications satellites from the Orbiter's payload bay. Originally selected as a scientist-astronaut by NASA in 1967, Joe served as mission controller (CAPCOM) for Apollo 15 and 17 and during Earth reentry of the first shuttle flight. A graduate of Depauw University and with a doctorate in physics from Yale University, Joe is author of numerous articles on physics, space operations, space research and science education. Author of *Entering Space: An Astronaut's Odyssey,* Joe has been an eloquent voice describing the experience of spaceflight.

Dr. Wendell Mendell has worked for NASA since 1963 and is currently Chief Scientist for Lunar Base Studies at the NASA Johnson Space Center. He has been an important strategist in the coordination of planning and development for future lunar missions leading to a manned Moon base. Long associated as a Planetary Scientist for the Solar System Exploration Division, Wendell participates in the development of long-range strategies for initiating and sustaining manned and unmanned planetary exploration. As an aerospace technologist for the Space Environment Division, Wendell was responsible for research on the thermal and photometric properties of the lunar surface. He has also served as a co-investigator and primary science team representative on the Apollo 17 infrared scanning radiometer experiment and conducted research on thermal emission spectroscopy of planetary surfaces. Wendell has authored numerous papers in scientific and popular journals and helped organize several conferences to promote a dialogue on long-range goals in space policy.

Notable Works in Biospherics

Space Biospheres
By Mark Nelson and John Allen, Introduction by Margret Augustine ($8.95)

Commencing with a concise integrative model of Earth's biosphere, the authors proceed to Biosphere 2, an extraordinary project in the creation of biospheric system to further humankind's ability to live in harmony with the sphere of life — on Earth or among the stars.

> "*Space Biospheres ... may help bring about the overdue realization that biological systems are essential to humanity's activities in space ... Anyone interested in the human settlement of space will find this a fascinating and useful addition to their library.*" — Gerard O'Neill, **New Scientist**, London

The Biosphere Catalogue
Edited by Tango Parrish Snyder ($12.95)

The Biosphere Catalogue pulls together state of the art information by leading scientists and thinkers in the various spheres of the biosphere — biomes, atmosphere, geosphere, evolution, communication, analytics, cities, travel, and more.

The Biosphere
By Vladimir Vernadsky ($5.95)

The first English publication of the original theory of the biosphere written in 1926. Preface by Dr. Evgenii Shepelev from the Institute of Biomedical Problems, Moscow.

Where The Gods Reign:
Plants and Peoples of the Colombian Amazon

By Richard Evans Schultes ($20.00)

Journey the mysterious Colombian Amazon with this most remarkable scientist and eloquent guide. With 140 black and white photographs from 1940-1954, Dr. Schultes presents an exquisite and moving portrait of his historic studies of ethnobotany in this remote region.

White Gold
The Diary of a Rubber Cutter in the Amazon 1906-1916

By John C. Yungjohann • Edited by Ghillean T. Prance ($7.95)

The story of one man's amazing struggle for survival as a rubber tapper in Brazil at the turn of the century — the height of the rubber boom. Dr. Ghillean T. Prance, Director of the Royal Botanic Gardens at Kew, introduces the book with a narrative on the present day efforts of the rubber cutters union to preserve their way of life in the Amazonian rainforest, and a comparison between the Amazon rainforest at the turn of the century and today.

Traces of Bygone Biospheres
By Andrey Lapo ($9.95)

The role of life in geological processes is revealed via the study of biomineralization and biosedimentation phenomenon — the traces of bygone biospheres.

Synergetic Press
Post Office 689 • Oracle, Arizona 85623 • (602) 896- 2920 • FAX (602) 896-2027